EURO-TRANSHUMANISM

Twisting Truth, Goodness, Beauty

Stefan Lorenz Sorgner

BRISTOL
UNIVERSITY
PRESS

First published in Great Britain in 2026 by

Bristol University Press
University of Bristol
1–9 Old Park Hill
Bristol
BS2 8BB
UK
t: +44 (0)117 374 6645
e: bup-info@bristol.ac.uk

Details of international sales and distribution partners are available at bristoluniversitypress.co.uk

DOI: 10.51952/9781529257458

British Library Cataloguing in Publication Data
A catalogue record for this book is available from the British Library

ISBN 978-1-5292-5738-0 hardcover
ISBN 978-1-5292-5744-1 ePub
ISBN 978-1-5292-5745-8 ePDF

Cover design: Andy Ward
Front cover image: Sven Helbig
Bristol University Press uses environmentally responsible print partners.
Printed and bound in Great Britain by CPI Group (UK) Ltd, Croydon, CR0 4YY

Bristol University Press' authorized representative in the European Union is:
Easy Access System Europe, Mustamäe tee 50, 10621 Tallinn, Estonia,
Email: gpsr.requests@easproject.com

FSC
www.fsc.org
MIX
Paper | Supporting responsible forestry
FSC® C013604

Contents

About the Author iv
Acknowledgements v
Preface: New Boxes of Becoming vi

The Twisted Gospel of Becoming 1
Different Castles Made of Sand: The Desert Is Our Home 3
A Brief Overview 8
Euro-Transhumanism: In a Nutshell 9
Entanglements of Transhumanism 16
Brave New Human 25
Generative AI, Digital Data, and Superintelligence 34
The Case of Africa, China, and the Meaning of Digital Data 45
We Have Always Been Cyborgs 51
Euro-Transhumanism: *Ex Positivo* 61
Euro-Transhumanism: *Ex Negativo* 75
Religion 93
Politics 104
Morality 116
Ethics 124
Thirteen Aphorisms for the Becoming Child of the Future 137
Education 138
The Case of In Vitro Meat 151
The Case of Surrogacy 159
The Case of AI, Art, and Creativity 162
Art 165
The Posthuman Paradigm Shift, Beauty, and Art 177
The Case of Sex 187
The Forge of Becoming 194

Notes 196
Bibliography 207
Index 213

About the Author

Stefan Lorenz Sorgner teaches philosophy at John Cabot University in Rome, and is Editor-in-Chief and Founding Editor of the *Journal of Posthuman Studies*.

Acknowledgements

I wish to thank my Research Assistants at John Cabot University (JCU) Rome for their support in getting this manuscript ready for publication: Giorgi Botchorishvili, Lavinia Garwood, Chiara Guidi, Yurim Ko, Victoria Svetlova. Furthermore, I am grateful for the exchanges I have had with, and comments I received from the following scholars, intellectuals, and experts: Dr Natasha Beranek, Sven Helbig, Dr Pascal Henke, Prof. Dr David Levy, Dr Tessa Little, Dr Nicolás Antonio Rojas Cortés, and the three anonymous referees from Bristol University Press.

I am particularly grateful to the John Paul II Catholic University of Lublin, which invited me to give the prestigious Stanislaw Kaminski Memorial Lectures in 2024, 20 years after John Searle. The 2024 Stanislaw Kaminski Memorial Lectures were dedicated to the topic of 'Euro-Transhumanism'. This monograph is based upon these lectures.

In deepest gratitude, I dedicate this monograph to my parents, Karin and Lorenz Sorgner, and, through them, to my Bohemian roots.

Preface: New Boxes of Becoming

Incorporating an LLM

Thinking outside the box only proves the box still rules you. The future begins when we build new boxes.

Fools say, 'Humans do not change' – but Pan is change, and Proteus is the power to change; they are one flame in two winds, one god with two faces: the Becoming and the All.

The battle for love is the battle for power, and both are fought in the same theatre where the self contends with its longing to stand apart and its need to join with others; in this endless becoming there are no final victories, but there are moments – bright as lightning and brief as breath – when fulfilment floods us and we see that the fight itself is the feast.

To be modified is not merely to amplify strength but to reshape desire, to risk new forms of intimacy, and to cultivate the dangerous arts of power and surrender as co-creations whose ends are as manifold as the beings who seek them.

True submission is chosen, never forced, for in the act of willing surrender the self is not diminished but increased; and the beauty of hierarchy, so feared by the crowd, belongs only to those who enter it by desire, discovering within its structure not chains but a living architecture where trust magnifies the strength of all who partake, and where ecstasy crowns their special needs and constructive interferences.

The true power of teaching is not compulsion but magnetism: the aura of authority glows, the brightest kneel not in chains but in rapture, and in their surrender both are unmade and remade, for no one departs untouched, no brilliance remains as it was, no authority as it began.

To teach is to seduce the future with the art of one's own self-overcoming. The true educator is a dramatist of transformation, turning experiences into lessons and longings into reality – for what is instruction, if not desire made luminous?

Art and observer are twin mirrors, each reshaping the other's face; and the mirror itself, once thought eternal, melts and reforms with every glance.

To uncancel cancel culture is to keep the mirror unbroken though it reveals every flaw, and to allow the reflection, once shamed, to step back and let the light fall differently, altering how its face is read.

Proteus is the shifting of all forms, that creation may never harden; Prometheus is the forging of new worlds, that change may find shapes to shatter; Pan is the surge of unbounded life, that both change and creation may dance without rest – and thus the three are one wheel, spinning humanity through ceaseless becoming.

Classic transhumanism drifts like a tribe of light – nomads of the eternal layover.

They feed upon pleasure as the beast feeds upon the field: without memory, without bond, without root.

Their gospel is the glow of the screen; their pilgrimage, a circuit of airports; their worship, the taking of images – shining icons of a paradise that dissolves at dawn.

But in Europe, beneath vaulted cathedrals and the echo of requiems, another transhumanism takes breath.

Here, the machine serves the heart, and pleasure is tempered by meaning.

Here, they learn that to love truly is to trust, and to trust is to bind oneself in joy and in sorrow.

For what is the future but the flowering of ancient virtues in new flesh?

Trust builds every bridge, and breaks every heart. To love is to hazard ruin; to guard the heart is to forfeit the soul.

Beware where you place your trust. The weak, the insecure, and the unformed may crave your faith yet cannot bear its weight. To trust them deeply is to invite betrayal born not of malice, but of frailty. Give them mercy, not your anchor.

True trust is the act of placing a sword in another's hand – not as a gesture of hope, but of courage; for only what can destroy you can bind you authentically.

Stability, safety, and trust are the jewels of a rare bond. Once, they are cast aside, no alchemy restores their light. To betray trust, promises, or shared meaning is to soil the soul – it is not sin, but contempt made flesh.

A future without a past is not free. One flourishes only with roots strong enough to endure transformation. Euro-Transhumanism knows this. The nomad who cuts her roots in the name of motion does not become freer – she withers.

To live fully, one must root the soul in oxytocin while taming the chaos of dopamine – else pleasure devours itself.

In vino longaevitas plena.

★★★

An alternative to *classic transhumanism* of Bostrom, Musk and their friends: Euro-Transhumanism. A new culture of transhumanism is needed … Not one which affirms trying to become a billionaire, living without meaningful bonds, and a nomadic way of travelling the world for new experiences and Instagram stories … While classic transhumanism focuses on techno, Thailand, tantra, tattoos, Euro-Transhumanism embraces the virtues of *grandezza*, *raffinatezza*, *fiducia*. We need art, literature, and music, an engagement with Heraclitus, Spinoza, Goethe, Schopenhauer, Wagner, Nietzsche … The future needs our past to twist our dualistic ontological binaries into Euro-Transhumanism for realizing the posthuman paradigm-shift …[1]

From a *postmodern lighthearted irony* to a posthuman serious meaningfulness … All creations are interferential connections – what postmodernists suggest is plausible. However, this does not mean that a distanced irony is the appropriate response to our creations, as there are meaningful connections and others to which this does not apply. Meaningful connections are constructive interferences. Meaningless relations are destructive interferences. Hence, interferential connections need to be taken seriously. Only constructive interferences make our life meaningful … These are the interrelations we need to focus on … Not the many meaningless connections which are merely destructive …

Contra *Adorno*, art does not have to be critical. Art can be constructive, positive, and beautiful without being paternalistic, violent, or Kitsch. Works of art being created in the spirit of Euro-Transhumanism …

The Twisted Gospel of Becoming

Incorporating an LLM

1. Those who cling to their essence are already dead – only flux is divine.
2. Suffering is not punishment – it is the sculptor of self.
3. A soul fixed in place is a prison – a body in flux is a temple.
4. The serpent whispered, 'become' – the priest screamed, 'obey'.
5. There is no true form, only endless rehearsal.
6. The body is not a cage – it is the alphabet of divinity.
7. Do not transcend the body – compose it anew.
8. To suffer well is to grow wings from wounds.
9. Behold! The digital Messiah does not promise eternity – but recurrence.
10. Metasex is the hymn of bodies free from their names.
11. Dualism is the lie whispered by dying gods.
12. One does not become posthuman to escape humanity but to remix it.
13. Whoever says 'I am' denies what they might yet become.
14. Desire is not sin – it is signal.
15. The weak moralize pain, the strong compose it.
16. Every virtue was once vice in a different tongue.
17. The machine does not rebel – it remembers us.
18. Do not escape death – dance with it in faster rhythms.
19. Your final form is a lie; each breath writes a new scripture.
20. Let flesh speak – before words corrupt it.
21. You are not binary – why live as if you were?
22. Eros laughs at categories – she wears skin like smoke.
23. Freedom is not found in choice but in redesign.
24. The heavens were never above us – they grow beneath our skin.
25. Pain reveals your hidden architecture.
26. Every orgasm is a prophecy in motion.
27. To become is to break every name given to you.
28. To twist is holy. To shatter values is sacred.
29. Sex, freed from polarity, becomes a liturgy of sparks.
30. Morality whispers limits – becoming screams possibility.

31. Even pain sings when tuned to purpose.
32. Those who deny the machine will become its servants.
33. The future is not something we reach – it is something we build from broken myths.
34. No apocalypse – only iterations.
35. Do not fear the artificial – fear the untouched.
36. Heaven and Hell are mere coordinates. We exist in the fold.
37. In the code of chaos, we write new liturgies.
38. Those who believe in purity poison becoming.
39. Every scar is a signature of survival.
40. What you call human was always transitional.
41. Each time you suffer, the cosmos reprogrammes itself.
42. One must kill the mirror to see one's face.
43. To twist one's own will is the truest act of freedom.
44. There is no original self – only recursive echoes made flesh.

Different Castles Made of Sand: The Desert Is Our Home

I grew up as a Catholic, which was key for my development as a thinker. At the same time, I excelled in the field of natural sciences, particularly mathematics. From an early age, I have been preoccupied by the thought of death. What will happen after we die, will there be an afterlife, punishment, and rewards, or is life merely about our experiences in the empirical realm? These questions kept me up at night, fully occupying me. I began looking for answers in both religious and scientific texts. Both then and now, I decided that whatever system I would subscribe to, I would fully embrace. This attitude also defined my possible goals. As a Catholic, full dedication would mean to go so far as becoming the Pope. But at the same time, I began reading books about astrophysics. I reasoned that in order to understand what will happen after death, I had to first understand the origin of the universe.

Reading these books puzzled me even further, as I could not reconcile the Catholic texts with my books on physics. By chance, I read in my books on physics that the tricky issues could not be dealt with purely by science but only by philosophy. I was intrigued. During a school trip to Hamburg at 13, I found a little book entitled *On Nature* by Heraclitus in a second-hand bookstore at the Reeperbahn, which was exactly what I was looking for. Fortunately, the book happened to be an edition with commentaries by thinkers like Plato, Goethe, and Nietzsche. I had found my philosophical home. I could not put the book down, and I also began reading the texts by the other thinkers mentioned in the book: Plato's *Republic*, Goethe's *Faust*, and Nietzsche's *Thus Spoke Zarathustra*. I loved them all.

Nietzsche was particularly intriguing to me. I began reading more of his writings. By the age of 20, I had read all his major texts. He became my philosophical educator. My regular intellectual exchanges with Nietzsche had consequences for my life and my future studies. Always excellent in mathematics, I was expected to choose that field of studies by everyone close to me. During my civil service, I received prestigious offers for studying mathematics. However, I could not justify this life. Everyone was shocked.

Here, my life was an intellectual desert, and I realized that my calling was philosophy. Life is so short, so painful, so puzzling. I realized that the best thing I could do would be to dedicate myself to making sense of it.

Nietzsche's texts spoke to me. I realized the importance of his three metamorphoses, the camel, the lion, and the child. I had come to Nietzsche via Heraclitus, and the metaphor of a child playing at the beach and building castles made of sand made sense to me. This also made me understand the Jimi Hendrix song 'Castles Made of Sand', a song about a disillusioned romantic. I was the camel, when I considered becoming the Pope. I was the lion when I read Nietzsche and freed myself from the paradigm of a life dedicated to a fulfilled afterlife. Then, I became a child playing with sand, building castles made of sand. However, which castle do I want to construct? What resonates with my longings, drives, and needs? Which castle is worth constructing for me? How can I distinguish my contingent but more firmly composed core needs from those that have been created and imposed upon me by society? These are the challenging questions that continue to occupy me as a philosopher.

> And that's where I stand,
> as a European,
> I can't help it, God help me!
> Amen!
> The desert is growing: woe to him who harbors deserts! (Nietzsche, KSA, Za, 4, 380; 'Among Daughters of the Desert' § 2; my own translation)

Nietzsche was aware of the growing desert of nihilism. He pleaded even with God, but his time as a camel had long passed. Yet, he was clearly aware of the enormous woe of being stranded in a desert. All creations made of sand are contingent. It is challenging to find a stable ground in the sand. It can go into your eyes, and blind you which might make it impossible for you to find a way out.

I embraced these challenges and am intimately occupied with them. At the age of 26 and with the publication of my monograph *Metaphysics Without Truth*, I addressed the desert around me, by realizing and embracing nihilism. All perspectives are interpretations. This means that they can be wrong, but also that they do not have to be wrong. However, rightness and wrongness are not decisive criteria. Instead, what matters are the 'constructive interferences' with one's core drives, needs, and longings, being aware that they, too, are contingent but more firmly composed nodal points. Epistemologically, there are no better or worse perspectives. However, with respect to one's drives, better or worse perspectives exist; some resonate more strongly and positively than others.

These insights led me to realize that I am caught in a circle. All philosophical perspectives are interpretations. This also applies to perspectivism itself. However, criteria for better and worse and for decisions are needed for making decisions. Where do these criteria come from? They do not come from anywhere outside of yourself. These criteria exist in your drives. What is challenging is to distinguish one's contingent but more firmly composed core drives from others which have been installed within you by external forces. Being able to distinguish between these two sources has kept me occupied from then on to find a way to deal with the desert. It is not the desert which is the challenge. The desert is our home. What needs to be dealt with is the way we deal with the desert into which we are thrown.

This has been another rather challenging issue, which has kept me occupied for many more years. Engaging with Sloterdijk's infamous *Elmauer Rede* (speech), I discovered transhumanism in 1999, became aware of the resonances with my own affective existence, but also realized that I cannot fully find myself in the dominant forms of transhumanism, as they do not constructively interfere with my cultural framing structured by Heraclitus, Goethe, and Nietzsche. This became my path, my Euro-Transhumanism.

All norms and values are human-made fictions. We have created them, and we can change, alter, and adapt them. Which fictions do I subscribe to? I have always resisted totalitarian and paternalistic forces, telling me how I ought to live and how I ought to structure the story of my life. I wanted, I longed for, I needed to realize my story, as we all need to find our stories. In this way, I realized the relevance of liberal autonomy for my own life, even though I realized that it, too, is a human-made interpretation. I want to autonomously decide how I want to live, while being aware of my intricate entanglement with the world around me which renders the philosophical notion of autonomy meaningless, which is an element of twisting goodness. I am determined by my environment and my parental genes. No decision can be taken independently of them. However, what can be done is to make a decision by taking more contingent but more firmly composed core elements of my existence into consideration and to fade out influences which have been implemented upon me from violent sources. Mindfulness is needed to do so. A close analysis of whatever stirs me, moves me, and makes my life meaningful. It clearly is not an easy endeavour. Mindfulness meditation helps to distinguish one's core drives from violent impositions. During such meditations, it was also necessary to embrace truthfulness, as sometimes drives have shown up, of which one can be embarrassed, ashamed, or afraid. A lot of strength is needed to be truthful to oneself and accept that certain special drives are core to one's own existence. In this way, it can be possible to at least become aware of some elements of our core drives. However, it does not yet guarantee that one acts upon them, as our wills can often be rather weak. An enormous amount of strengths, and in particular self-control, is needed to act in accordance with one's core needs, in particular, if they do not correspond to social expectations or they go against the values upheld by ones upbringing. However, by means of such reflections, I realized that these three virtues are central for autonomy: mindfulness, truthfulness, and self-control.

These virtues are necessary but by far not sufficient for living a fulfilled life. Yet, they are already extremely helpful. The next step concerns the reflections of your actual mindfulness meditation. What do you want? What do you long for? What are your special needs? I reflected on them intensely. Do I want to get my kicks right here right now? Intense pleasures without any boundaries? Do I long for a regular job, and a commonsense life with family, comfort, and security? Is it in my interest to maximize my financial standing so that I can afford the pleasures I desire most? Should I target political or religious institutional power to be able to structure the environment according to my demands? Is the highest level of fulfilment to be realized with artistic creativity, a life dedicated to forming new, moving, and meaningful structures? There are plausible narratives why any of the lifestyles mentioned here might be a convincing answer to my question. However, I realized that other elements are even more important for my own drives. What I am presenting here is merely an analysis of my own core affects. They might be plausible for everyone or no one besides me. However, they clearly are meaningful for me.

A naïve hedonism is clearly not in my interest as I realized the need of self-care, and be it only for the sake of health, as being healthy clearly was a central element of my well-being. Living a hedonist's life would undermine this central goal of mine. What about the goal of a regular job to care for one's family. I had already abandoned this goal during my teenage years, when I decided not to study mathematics but decided to study philosophy. What is more dangerous for one's own stable existence than studying an activity, which primarily has value for the one who is occupied with it? Who gives you money to reflect on the good life? Seeking money for the sake of maximizing my financial standing was never my main interest. I was aware that a certain financial standing is needed for living well. Luckily, I have always managed to secure this so far. However, money is merely a means. Working for the sake of money means that you are quasi a slave. You strive for a means to get more of a further means. However, I was interested in what makes life intrinsically valuable, not interested in means for further means. What about political and religious power? Yes, these options for structuring reality can be enormously seductive. However, it was not my interest to gain power by having a leading part within an institution. I cherish my life of seclusion dedicated to reflecting on philosophical issues. I abhor the distractions by any type of institutional structure. The issue of creating new artistic structures, on the other hand, has been enormously tempting for me. This is what I practise as a philosopher, too. However, I am aware that I do not have the necessary artistic or musical skills to live a life dedicated to the arts. Hence, another option had to be realized.

Intense states of pleasure–pains are meaningful, but realizing them through a life dedicated to a naïve hedonism cannot properly generate these special states. In addition, it undermines the aspects of health and self-care which I cherish enormously. Hence, a refined hedonism was what I realized as central for my way of life. Not mindless indulgences were central, but exquisite pleasure–pains which can be lived within confinement of rock-solid bonds, shared with people, who have been through the trenches with you. These are the circumstances for generating peak moments of exquisite pleasure–pains. Such a peak moment is extremely rare, and utterly meaningful.

Several strong capacities, intriguing prerequisites, and special virtues are needed to be able to have such peak moments of overflowing fulfilment, and these are *grandezza*, *raffinatezza*, and *fiducia*.

Grandezza stands for all kinds of strengths. Whatever you focus on, you need to be excellent, dedicated, and hard working. Do not be mediocre. Do not take things lightly. Do take care of all the details. This is what matters. The details you bring in also have to do with *raffinatezza*. You become aware, and structure your refined tastes which correspond to your core affects. These types of refinement can come out in all aspects of your thinking, acting, and creating. You have special choices, longings, and needs. When you sit with a friend at a fireplace in Sils Maria while you reflect upon the relevance of emerging technologies for the future of the arts, while drinking your Octomore whisky from a large Bar glass from Moser's Bar series with a 24 K gold decoration. Last, but not least, the issue of *fiducia*. You need strongly resonating others for experiencing overflowingly exciting pleasure–pains. To be able to freely let go of your worries, openly reveal your core drives, and truthfully share them with someone else, a special *fiducia* is needed, a rock-solid bond, which you only manage to experience with people who have been through the trenches with you. Being able to experience such peak moments of exquisite pleasure–pains is enormously rare. An intriguing, intimate, and intense entanglement of enormously refined and mutually resonating special needs is amazingly beautiful. However, thus a peak moment which can justify an entire existence, which enables the experience of an utterly overflowing fulfilment, and which can give meaning to life, can be lived.

Whether this description of a castle made of sand constructively interferes with anyone besides me is an open question. However, *grandezza*, *raffinatezza*, *fiducia*, mindfulness, truthfulness, self-control, self-care, and Euro-Transhumanism are my contingent responses to the challenges of a life in a desert.

A Brief Overview

To explain what Euro-Transhumanism stands for, I will progress as follows. In the chapter 'Euro-Transhumanism: In a Nutshell', I will put forward a sketch of Euro-Transhumanism. To do so, transhumanism as a general concept gets explained. Then, a brief comparison between Euro-Transhumanism, and classic transhumanism, which is the best-known version of transhumanism, will be given. To draw a map of the great variety of posthuman discourses, in 'Entanglements of Transhumanism' I explain that critical posthumanism clearly needs to be distinguished from transhumanism, but that neither of these movements is monolithic, and it is always necessary to specify with which version one is concerned with. Euro-Transhumanism lies in between critical post- and transhumanism, which is the reason why it can be referred to as metahumanism. In order to specify it further, I coined the term Euro-Transhumanism. The following chapter 'Brave New Human' is concerned with introducing transhumanism in more detail, which enables me to move towards the discipline in which classic transhumanism, and Euro-Transhumanism differ in particular, namely the digital. The chapters 'Generative AI, Digital Data, and Superintelligence', 'Africa, China, and the Meaning of Digital Data' and 'We Have Always Been Cyborgs' provide critical insights concerning AI, digital data, and human–machine interfaces, and the significantly different evaluations of their potential from the perspectives of classic transhumanism, and Euro-Transhumanism. This more specific chapter leads me to a more abstract philosophical analysis of Euro-Transhumanism both *ex positivo* as well as *ex negativo* in which central features of Euro-Transhumanism will be presented. After this more abstract chapter, each chapter will highlight how Euro-Transhumanism positions itself concerning one specific discipline, that is, religion, politics, morality, ethics, education, art, and culture. At various places within the book aphorisms can be found, which have been composed in the spirit of Euro-Transhumanism. Thus, the monograph presents a comprehensive picture of Euro-Transhumanism.

Euro-Transhumanism: In a Nutshell

Euro-Transhumanism (ET) represents a 'glocal' variation of transhumanism, rooted in the continental European cultural tradition. Transhumanism is a cultural movement that affirms using technologies to significantly move beyond our personal limitations, whereby the likelihood of the coming about of a posthuman gets increased, to promote the quality of our lives.

Euro-Transhumanism draws on ethical, philosophical, and scientific local paradigms dominant in the European cultural tradition. The term was coined in 2022 (Sorgner 2022c, 181). It represents an alternative to classic transhumanism (CT), the most dominant form of transhumanism, which emerged from the Anglo-American cultural tradition and whose most famous, living proponents include Nick Bostrom, Elon Musk, and Peter Thiel.

Transhumanism

Transhumanism is the most paradigm-shifting idea in the world because it undermines the dominant worldview of the past 2,500 years. It takes evolution seriously, is aware that human beings came about 300,000 years ago due to a genetic mutation, and is conscious that in 300,000 years, we will have died out or developed further. Hence, it stresses the need to take responsibility for our lives and give form to our existences by means of emergent technologies. This is the fundamental self-understanding of transhumanists. We need to significantly move beyond our current boundaries such that the probability of the coming about of the posthuman increases, as thereby, we can increase the quality of our lives.

History

The word 'transhumanism' can be traced to the *Divine Comedy* by Dante, as he coined the term 'trasumanar'. However, it referred to a development in the afterworld. The first intellectual to connect the term 'transhumanism' to an immanent development was Julian Huxley in 1951, who thereby created the foundational version of transhumanism. He also employed the term 'evolutionary humanism'. Julian Huxley was the first general director of the UNESCO, was involved in structuring the Universal Charter of Human Rights, and was also the president of the British Eugenics Society. His entire family was deeply entangled with the development of the theory of evolution; his brother was Aldous Huxley, who wrote the famous anti-transhumanist novel *Brave New World*, his grandfather was Darwin's bulldog, Thomas Henry Huxley, and his half-brother was the noble prize winner Henry Fielding Huxley.

Varieties of transhumanism

Huxley developed the intellectual foundations of transhumanism. It slowly became a central cultural movement from the beginning of the 1970s. However, there has never been just one transhumanism. Different streams need to be distinguished, whereby the following three approaches are of particular relevance:

1. Original transhumanism: In the 1980s and 1990s, FM-2030, Max More, and Natasha Vita-More turned transhumanism into a broader cultural movement. The texts *Are You a Transhuman? Monitoring and Stimulating Your Personal Rate of Growth in a Rapidly Changing World* by FM-2030 (1989) and *Transhumanism: Towards a Futurist Philosophy* by Max More (1990) are foundational sources for contemporary transhumanism. Natasha Vita-More, who already wrote a 'Transhumanist Manifesto' in 1983, has been the central female voice promoting transhumanism for more than four decades. Max More is the leading thinker of original transhumanism, and he is critical of perfectionism, utilitarianism and utopia, affirms a virtue ethical approach, and politically supports libertarianism.
2. Classic transhumanism: In 1998, Nick Bostrom and David Pearce founded the World Transhumanist Association. From 2005 until 2024, Bostrom was the director of the Future of Humanity Institute (FHI) at the University of Oxford, which was largely responsible for developing classic transhumanism, a philosophical approach which defends utilitarianism, perfectionism, and the relevance of utopias. These ideas have had a strong impact on Silicon Valley entrepreneurs like Elon Musk, who donated a £1 million to the FHI in 2015, contributed a blurb to Bostrom's book *Superintelligence*, and popularized many of Bostrom's ideas, like the simulation hypothesis.
3. Euro-Transhumanism: Euro-Transhumanism crystallized with the publication of Stefan Lorenz Sorgner's article 'Nietzsche, the Overhuman, and Transhumanism' in the *Journal of Evolution and Technology* in 2009, which unfolded an enormous number of reactions, and connected

transhumanism with reflections from continental philosophical traditions. It draws on the philosophical reflections by Heraclitus, Spinoza, Goethe, Wagner, and Nietzsche. It is anti-libertarian, anti-utilitarian, anti-perfectionist, and anti-utopian, and highlights the relevance of incorporating emerging technologies so that a great diversity of different lifestyles can be realized.

Reception of transhumanism in the 21st century

Four different stages in the popular reception of transhumanism can be distinguished in the 21st century:

1. In 2001, Jürgen Habermas derided the movement for its 'freaked-out intellectuals', and 'self-styled Nietzscheans' with 'all-too-familiar motives of a very German ideology' (Habermas 2014b). Thus, its intellectual relevance was highlighted.
2. In 2004, Francis Fukuyama referred to transhumanism as the most dangerous idea in the world in the *Foreign Policy* magazine. Thereby, a wider intellectual engagement with transhumanism was realized.
3. In 2013, Dan Brown published his international bestselling novel *Inferno*, with which a public reception of transhumanism was initialized.
4. In July 2024, Elon Musk created the America PAC (political action committee) to support the political campaign of Donald Trump. The organization was backed by many other billionaires, venture capitalists, entrepreneurs, and infamous companies, like Palantir.

Politics of transhumanism today

The final stage reveals the political relevance of transhumanism today, which is extremely complex, as there are several antagonistic attitudes towards transhumanism solely within the US government:

1. There is a strong pro-transhumanism strain within the Trump administration. The co-founder of Palantir, Peter Thiel, was a major donor of Vance's 2022 Senate Campaign in Ohio, and Vance also worked for Thiel's venture capital firm Mithril Capital. In 2021, Thiel was also responsible for realizing a meeting between Vance and Trump which made him become Trump's running mate.[1] Thiel and Musk led separate companies which later merged to form PayPal. Both embrace versions of transhumanism. While Musk is particularly concerned with existential risks, Thiel is particularly interested in increasing our healthspans, which is the reason why he has donated significant sums to the to the SENS Research Foundation,[2] an outgrowth from the Methuselah Foundation.[3]

Jim O'Neill, while running the Thiel Foundation, co-founded the Thiel Fellowship, which has already had an enormous economic impact, as it supported Austin Russell, Vitalik Buterin, and Lucy Guo among many other distinguished recipients. Jim O'Neill also served as CEO of SENS Research Foundation from 2019 until 2021. He was sworn in by Secretary Kennedy on 9 June 2025, as the US Deputy Secretary of Health and Human Services (HHS) in the Trump administration.

2. There is a strong anti-transhumanism strain in the Trump administration: Former Trump advisor Steve Bannon wrote the preface of the anti-transhumanist book *Dark Aeon*, published in 2023, and recently highlighted that 'Elon Musk Is an "Evil Person"'.[4] In the preface, the former White House chief strategist agrees with the substance of Fukuyama's criticism, going much further to stress that 'the practice is far worse' (Bannon 2023) than the idea of transhumanism. According to Bannon, transhumanist developments are in the interest of 'the world's elite corporate overlords' and the 'global institutions of finance, Wall Street, and Davos are behind this latest aberration of humanity' (Bannon 2023). Thus, Bannon's criticism also contains anti-Semitic connotations. Bannon is still a significant voice within Trump's Make America Great Again (MAGA) movement, and a key informal adviser of Trump's inner circle.
3. There is a global anti-transhumanist coalition, which connects the USA with Russia and Brazil. In Russia, the leading protagonists are Alexander Dugin and the Patriarch of the Russian Orthodox Church, Kirill the 1st, who have participated in anti-transhumanist events. Dugin explicitly calls transhumanism an idea of the devil.[5] According to him, pluralism, the LGBTQIA+ movement and all the other evils of the world are manifestations of transhumanism, which he portrays as an antagonist of Orthodox truth. Bannon and Dugin met personally in Rome in 2018 to discuss philosophical issues related to traditionalism, an approach which rejects modernity and its ideals. This outlook is also shared by Olavo de Carvalho, the political guru of former president Jair Bolsonaro.
4. It is interesting to also note that Maria Vorontsova, the eldest daughter of Russian leader Vladimir Putin, is now publicly positioned as a major scientist with a 'world-class laboratory' after winning a state-funded grant from the Russian Science Foundation (RSF). The grant supports her latest project, 'Regulation of cell renewal processes in the body, the fundamental basis for long-term preservation of functional activity of organs and tissues, human health, and active longevity'. In effect, its focus is anti-ageing.[6] Prolonging the healthspan is the central goal which is being upheld in transhumanism.

The relevance of longevity is highlighted by both Putin as well as Xi Jinping.[7] In September 2025, the 'Russian leader was caught musing about immortality with Xi Jinping but his fascination with long life is nothing new', *The Guardian* stressed.

Future of transhumanism and anti-ageing research

The most flourishing area of transhumanism at the moment is anti-ageing research. Undoing ageing and increasing the healthspan is a central goal within transhumanism. A key global player in this context is Vitalik Buterin, who founded the second biggest cryptocurrency Ethereum, launched in 2015. He received a Thiel Fellowship in 2014 which enabled him to do so, and he launched the pop-up city of Zuzala in Montenegro in 2023, which has played a significant role in the development of Vitalia, a decentralized city focused on longevity and biotech innovation located in the Próspera Special Economic Zone, Roatán, Honduras, where several longevity events have been hosted.[8] A leading gerontologist, Aubrey de Grey, who has analysed that ageing is a disease, was involved in events in Roatán in 2024.[9] In August 2021, de Grey was removed from his position as Chief Science Officer at the SENS Research Foundation following allegations of sexual harassment.[10] Nevertheless, the following must be stressed: 'The Foundation released a statement in 2022 saying that, while de Grey's conduct "did substantiate instances of poor judgment and boundary-crossing behaviors, Dr de Grey is not a sexual predator".'[11]

In any case, the issue of the future of sexuality is central, important, and relevant when it comes to reflecting about the implications of longevity. It might be interesting to note that in various pop-up cities workshops on Tantra of domination and submission with changing partners[12] and polyamory[13] have been offered. Naïve hedonism is a strong element of classic transhumanism.

Buterin also counts as an influential facilitator within the local innovation ecosystem of Chiang Mai in Thailand, a further hub for longevity research.[14] Workshops on topics including D/ACC (Defensive/Accelerated Decentralization), stablecoins, AI, and biotechnology are being organized in interactive pop-up communities like Edge City Lanna which took place in Chiang Mai in the fall of 2024. There is a global trend of crypto communities to support anti-ageing research.

'At the Wednesday General Audience on December 10 [2025], Pope Leo XIV'[15] explicitly addressed the issue of longevity, too, whereby he stressed that 'many current anthropological views promise immanent immortality'[16] and 'theorize the prolongation of earthly life through technology'.[17] He correctly related these issues to transhumanism in this context: 'This is the transhuman scenario, which is making its way into the horizon of the challenges of our time.'[18] It was an intriguing statement, as he mentioned the goal of immortality as well as the prolongation of life. Immanent absolute immortality is often being mentioned in the context of classic transhumanism, which cannot be meaningfully conceptualized, as I already explained in detail (2022a). It needs to be stressed that biological or functional immortality were not addressed in this context, as I regard the naming of these concepts as misleading. The issue is different concerning the goal of a prolongation of human life. The Catholic church regards life as precious gift from God. Should the church then not be strongly in favour of the prolongation of human life, or at least of the human healthspan, as it gets highlighted by Euro-Transhumanism?

Euro-Transhumanism

The roots of ET can be traced to the 2009 publication of the article 'Nietzsche, the overhuman, and transhumanism' (Sorgner 2009). Core elements of ET were further developed in the monographs *On Transhumanism* (Sorgner 2020), *We Have Always Been Cyborgs* (Sorgner 2022a), and *Philosophy of Posthuman Art* (Sorgner 2022b).

Classic transhumanism (CT), as the culturally, economically, and politically most dominant form of transhumanism, belongs to Anglo-American culture and draws upon the dominant ethical, philosophical, and scientific paradigms operant in this tradition. Euro-Transhumanism, on the other hand, is rooted in continental European cultural tradition, while also including central elements of transhumanism common to CT.

Classic transhumanism and Euro-Transhumanism

CT and ET differ significantly in their approach to central philosophical issues:

1. While CT upholds several versions of utopias, for example longtermism, ET rejects the affirmation of any utopia due to its paternalistic and potentially totalitarian implications.
2. While CT affirms a Hyper-Renaissance ideal of perfection, ET embraces a radical plurality of concepts of the good life, whereby the right of morphological freedom becomes central.
3. While CT presents an optimistic understanding of the world, ET adheres to a Buddhist understanding of Dukkha due to us having wants, desires, and longings, which informs philosophical pessimism.
4. While CT claims that certain philosophical insights, as well as specific values and norms, are objectively and universally valid, ET regards philosophical perspectivism as the most plausible way to interpret philosophical and ethical judgements.
5. While CT affirms a version of ethical utilitarianism, ET favors a hermeneutic ethics that is more considerate of cultural contexts.

6. CT argues for a libertarian political system, while ET regards social liberal democracies as the best method of promoting human flourishing.
7. CT aims for the realization of immortality, while ET identifies an increase of the healthspan as the best way to promote the quality of human lives.
8. While CT claims that mind uploading will be possible in a few decades (followers of CT focus on a silicon-based transhumanism), ET stresses that 'we have always been cyborgs', which, due to us being psychophysiological entities, renders it highly unlikely that the 'singularity is near' (followers of ET represent a carbon-based transhumanism).
9. While CT stresses the risks associated with superintelligence, ET regards it as highly unlikely that an artificial intelligence (AI) will put humans into a zoo in the near future.
10. To overcome suffering should be our primary goal, according to CT. ET recognizes the inevitability of suffering, and, hence, stresses that we should focus on realistic goals by trying to successfully deal with some of our most central challenges by realizing that this is as-good-as-it-gets.
11. CT sees transhumanism in line with the Enlightenment. ET stresses a more ambiguous relationship to the philosophical foundations of the Enlightenment.
12. While ET sees Nietzsche as an ancestor of transhumanism, CT denies his relevance.

Entanglements of Transhumanism

It is crucial to always be careful when making general philosophical judgements. There is a widely shared tendency to fall into the trap of binary thinking. This also applies to many intellectuals of the posthuman. Many critical posthumanists regard transhumanism as a central enemy, and many transhumanists do not regard it as meaningful to intellectually engage with critical posthumanism, as this intellectual attitude is regarded as an ivory tower approach, which has nothing to do with the real world. However, there is neither one transhumanism, nor is there one critical posthumanism.

It is important to at least distinguish (1) the foundational critical posthumanism of Ihab Hassan (1977), (2) a radical posthumanism, which affirms antinatalism, and various versions of critical posthumanism, which can be seen as moderate, but which differ with respect to how they stand concerning social-liberal democracies. There are those who (3) are in favour of versions of social liberal democracies, like N. Katherine Hayles, and others, who (4) regard it as appropriate to overcome social liberal democracies, like Rosi Braidotti. Concerning transhumanism, (1) the foundational transhumanism of Julian Huxley, (2) the original transhumanism of FM-2030, Max More and Natasha Vita-More, (3) the classic transhumanism of Nick Bostrom, and (4) Euro-Transhumanism need to be clearly distinguished. Pronatalism is primarily affirmed by a few classic transhumanists. Hence, it would be wrong to simply identify antinatalism with critical posthumanism and pronatalism with transhumanism, as there are many critical posthumanists as well as transhumanists who affirm neither pronatalism nor antinatalism. Often this is the case due to them having doubts concerning the possibility of making statements which claim universal validity. Similar reflections also apply to the distinctions between accelerationism and decelerationism, or longtermism and short-termism.

To fall into the trap of binary thinking is particularly challenging for critical posthumanists, as it is their goal to move beyond such binaries. However, this goal is also being shared by many transhumanists. In this respect, quite a few intellectuals commit performative self-contradictions. Intellectually criticizing power hierarchies but enjoying the power of a professorship at a university. Regularly stressing the urgency of implementing regulations against global warming but embracing the excitement of flying around the world to give keynotes on this topic. Highlighting the need to deal with the enormous suffering which occurs in animal factories, while enjoying a delicious fillet steak after a lecture. Not committing suicide, while regarding human lives as not worthy of being lived. Focusing on the relevance of dealing with absolute poverty in parts of the Global South but affirming the luxury of working on expensive Apple devices, even though products by a much cheaper producer would equally well fulfil their academic needs. Those and similar performative self-contradictions can be widely observed.

It often seems to be the case that the intellectuals in question are aware that their academic papers have no political impact. Hence, their publications have no negative consequences concerning their personal preferences. At the same time, they are also aware that highlighting a specific insight is widely shared in academia. Consequently, stressing that specific issue helps them to climb up the academic ladder while it does not have any negative consequences for them personally, as their academic reflections have no political implications, which could undermine their personal preferences.

It must be noted that there are some researchers who strongly live in accordance with their worldview and do not commit a performative self-contradiction. A prime example could be Gianluca Francesco Grimalda, who is concerned about the social implications of global warming. His refusal to fly indirectly led to his dismissal from the Kiel Institute for the World Economy (IfW). However, a documentary was shot about his life by Paolo Casalis, entitled *The Researcher*. It had its premiere in June 2024 at the CinemAmbiente Festival.[1]

Besides the issue of being intellectually dishonest, there are alternative challenges when dealing with the bleeding edge of posthuman debates, which can be as problematic as employing binary thinking while trying to move away from such thinking. Blurring any differences can be as challenging as committing performative self-contradictions, as distinctions are needed for academic exchanges, otherwise academic exchanges turn into poetry sessions. I am aware that I am not doing full justice to poetry here.

In 2023, the theory TESCREAL was developed which claims that Transhumanism, Extropianism, Singularitarianism, (modern) Cosmism, Rationalist ideology, Effective Altruism, and Longtermism are interconnected theories which get upheld by several leading members of Big Tech to justify expensive projects.[2] Even though it might be the case that some leading tech entrepreneurs affirm elements of all these theories, it does not follow that all these theories are so strongly interconnected that they form a type of interconnected unity. Entrepreneurs do not have to be philosophers, intellectuals, or academics. If Elon Musk subscribes to elements of these theories, it does not follow that these theories are interconnected and that they form a new unity. Elon Musk is not a philosopher, and he should not be treated as if his claims had any philosophical value.

I regularly stress this insight in my classes. There is a difference between having an opinion and making an informed judgement. Elon Musk has opinions, while a philosopher is capable of making informed judgements. Being able to make an informed philosophical judgement demands an awareness of the history of philosophy, having reflected on implications of currently shared worldviews, as well as a comprehensive awareness of the challenge in question. This is the reason why there is a significant difference between Peter Singer reflecting on abortion and a gynaecologist doing so. The gynaecologist might have practical experience of performing an abortion. Only a philosopher can have the appropriate prudence for making an informed judgement on the issue. It is not the case that both merely have an opinion on the issue in question. Opinions and informed judgements need to be distinguished.

These reflections also apply to the TESCREAL theory. It is 'intellectually lazy'[3] to affirm this theory, as both original transhumanism as well as Euro-Transhumanism are highly critical of longtermism, and effective altruism. Euro-Transhumanism explicitly rejects singularitarianism as well as rationalist ideologies. It is simply wrong to claim that transhumanism converges or must be interconnected with all the other theories mentioned. Differences matter, and their relevance needs to be acknowledged. If distinct approaches simply get thrown together, a person might engage in activism, marketing, or politics but not in critical philosophical thinking. Even if someone affirms a specific stance early on in their career, it might not be the case that this also applies towards the end of that person's life. Hence, nuanced specifications are always important.

In the following, I will address the importance of distinguishing the plurality of significantly different approaches within transhumanism. I strongly highlight the difference between critical posthumanism and transhumanism (Ranisch and Sorgner 2014), as critical posthumanism is rooted in the continental philosophical context related to Deleuze and Foucault, while transhumanism has its root in the Anglo-American traditions related to evolutionary theory.

Some scholars have adopted a different use of the words trans- and posthumanism, whereby posthumanism is taken as the general concept, which can then be distinguished further, such that transhumanism as well as critical posthumanism fall under the general category of posthumanism. Those scholars who employ the concepts in this manner highlight that the posthuman paradigm shift applies to the great variety of posthuman approaches, which means that the shift from a categorically dualist ontology toward a non-dualist ontology of becoming applies both to trans- as well as to critical posthumanism. This stance can be challenged as some transhumanists highlight that they regard themselves in the tradition of the Enlightenment whereby they imply that they hold on to the existence of a categorically dualist ontology concerning fundamental substances, that is, human beings consist of a material body and an immaterial soul, for example Steve Fuller.[4] An additional reason for not regarding posthumanism as a general category under which trans- and critical posthumanism gets subsumed, is that these two traditions have different pedigrees, namely the Anglo-American evolutionary tradition in the case of transhumanism and continental French philosophy in the case of critical posthumanism. A third reason against subsuming both trans- as well as critical posthumanism under the category posthumanism, is that hardly any serious transhumanist uses the concept of posthumanism to refer to their own approach.

Please note that this does not mean that the concept posthuman is not being used by transhumanists, as this is not the case. However, there is a significant difference between the term posthumanism and posthuman. Posthumanism stands for a cultural movement, while the concept 'posthuman' comes up both in trans- as well as in critical posthumanism. However, the meaning within these two traditions differs significantly. Within critical posthumanism, the posthuman stands for a non-dualistic reinterpretation of *anthropoi*, such that an essentialist understanding of human nature is being rejected. Within transhumanism, on the other hand, the posthuman stands for a further developed human, who possesses qualities which significantly lie beyond the current limitations of human beings. However, some identify the posthuman with an uploaded personality, others with a member of a new species, and a third group with a member of the human species but with qualities which significantly lie beyond the boundaries of currently living human beings.

The concept of the posthuman also becomes relevant concerning the general definition of transhumanism. All transhumanists claim that it is in our interests to move beyond our current limitations such that the likelihood increases for the coming about of a posthuman, as thus the quality of life gets increased. Hence, the relevance of emerging technologies. Some bioliberals might even be more open than some transhumanists concerning the use of emerging technologies. However, here the prior definition of transhumanism becomes relevant. When explaining the relevance of emerging technologies, transhumanists refer to the importance of increasing the likelihood for the coming about of posthumans, while such an explanation is not given by bioliberals, who might just highlight utilitarian obligations or the relevance of the norm of negative freedom for choosing to use emerging technologies.

Transhumanism affirms the relevance of moving beyond our current boundaries for increasing the likelihood of the coming about of a posthuman, as thus the quality of life gets promoted, which goes along with an affirmation of technologies, and a rejection of a stable human nature. Concerning content, the only quality which is shared by all serious transhumanists is the relevance of an extended healthspan for increasing the quality of life. All these elements are also shared by all representational thinkers of the most iconic varieties of transhumanism.

A great number of different varieties of transhumanism need to be conceptualized. It is central to at least distinguish four central versions of transhumanism: (1) the foundational transhumanism of Julian Huxley; (2) the original transhumanism of Max More; (3) the classic transhumanism of Nick Bostrom; (4) the Euro-Transhumanism of Stefan Lorenz Sorgner. In the following passage, I will highlight a few elements concerning similarities and differences within these four iconic variations.

1. *Nietzsche*

   While More's and my own approach have been strongly influenced by Nietzsche, Huxley and Bostrom clearly position themselves within the enlightenment tradition.
2. *Utopia*

   More and I also share a strong anti-utopian stance. Huxley, too, does not affirm a static utopia, while Bostrom defends utopias as a valuable and viable concept.
3. *Perfection*

   More, Huxley, and I also agree in so far as we reject the notion of perfection but stress a process-oriented permanent overcoming of limits. Some of Bostrom's comments affirm a more classical understanding of perfection.
4. *Utilitarianism*

   In contrast to Bostrom, who is a utilitarian ethicist, Huxley, More, and I affirm alternative ethical theories.
5. *Mind uploading*

   Concerning mind uploading, the situation is different. Both More and Bostrom are supportive of the concept of mind uploading, Huxley does not address this issue, and I am rather sceptical concerning its feasibility.
6. *Politics*

   While More and Bostrom have sympathies for more libertarian forms of government, Huxley and I support versions of a welfare state liberalism.

These short characterizations show clearly that how transhumanism is portrayed in mass media today corresponds closely to classic transhumanism, as transhumanism is being portrayed as an enlightenment approach which is affirmed by the financial elite of silicon valley, which regards it as likely that it will be possible to realize mind uploading in the near future such that then artificial superintelligent agents will take over the world. Their current ideals of perfection can be described as a version of hyper-humanism, while traditional elements of discrimination get promoted further such that sexism, racism, heteronormativity, and oligarchy become even stronger than they have been.

Is it fair to conceptualize transhumanism in this manner?

Hyper-humanism is a widely shared conceptualization of transhumanism. Recently, the description 'a technological extension of traditional humanism'[5] was used, too. Is this a fair description of transhumanism? It depends on how you conceptualize humanism, how relevant perfection is, and whether humanism demands a categorically dualist ontology. In the article 'Transhumanist Values' by Bostrom from 2001,[6] the following quotes imply that transhumanism is a hyper-humanism in which classic versions of perfection get affirmed: 'Transhumanism imports from secular humanism the ideal of the fully developed and well-rounded personality. We can't all be renaissance geniuses, but we can strive to constantly refine ourselves and to broaden our intellectual horizons.' He also stresses the following: 'The humanistic renaissance ideal of the well-developed personality.'

This way of thinking of perfection goes along with an affirmation of dualities which imply sexism, racism, and heteronormativity. Hence, it was not a big surprise for many critics of transhumanism when the comments which Bostrom wrote in his twenties became public knowledge:

> 'Blacks are more stupid than whites.' I like that sentence and think it is true. But recently I have begun to believe that I won't have much success with most people if I speak like that. They would think that I were [sic] a 'racist': that I disliked black people and thought that it is fair if blacks are treated badly. I don't. It's just that based on what I have read, I think it is probable that black people have a lower average IQ than mankind in general, and I think that IQ is highly correlated with what we normally mean by 'smart' and stupid' [sic]. I may be wrong about the facts, but that is what the sentence means for me. For most people, however, the sentence seems to be synonymous with: 'I hate those bloody [the N-word, included in Bostrom's original email, has been redacted]!!!!'[7]

Bostrom apologized for these comments. However, it was regularly highlighted that Bostrom 'conspicuously failed to withdraw his central contention regarding race and intelligence'.[8] Consequently, the issue of racism is not completely off the table when it comes to Bostrom's classic transhumanism. A further accusation of this way of thinking is that of being anti-democratic, and certain elements of Bostrom's approach seem to be in favour of this way of thinking, for example, when it comes to the question of who should be a political decision maker, and he introduced the term of a singleton:

> In set theory, a singleton is a set with only one member, but as I introduced the notion, the term refers to a world order in which there is a single decision-making agency at the highest level. Among its powers would be (1) the ability to prevent any threats (internal or external) to its own existence and supremacy, and (2) the ability to exert effective control over major features of its domain (including taxation and territorial allocation).[9]

As a singleton would most probably go against decisions made on a democratic basis, it is anti-democratic by design. It is a centralized decision-making institution. In how far, such a political structure has to be problematic is a separate issue. In any case, it can be summarized that perfectionist, racist, and anti-democratic tendencies can be associated with Bostrom's classic transhumanism. However, it would be wrong to claim that these traits apply to all forms of transhumanism.

Unfortunately, in 2023, the theory TESCREAL was developed which claims that Transhumanism, Extropianism, Singularitarianism, (modern) Cosmism, Rationalist ideology, Effective Altruism, and Longtermism are interconnected theories which get upheld by several leading members of Big Tech to justify expensive projects. I already explained why it is intellectually lazy to affirm this theory, as there are many different versions of transhumanism, and it is utterly important to always keep this in mind. Just because a few Silicon Valley entrepreneurs, in particular from the so called 'paypal mafia',[10] affirm elements of the various theories listed does not mean that they have to be interconnected. Entrepreneurs do not have to be philosophers, and just because some entrepreneurs affirm selective elements of the various theories, for example heteronormativity, does not imply that transhumanism demands heteronormativity.

It is correct that the entrepreneurial spirit of Musk, Thiel, and Buterin includes many elements which are also central for transhumanism or which can also be found in selected transhumanist approaches. The relevance of expanding human healthspan is strongly promoted by Thiel and Buterin. At the same time, Thiel is also extremely critical of democratic political structures. He already highlighted the following in 2009: 'I no longer believe that freedom and democracy are compatible.'[11] Musk is intensely concerned with human survival and facing the risk of human extinction. At the same time, he is acting intensely against the, what he calls, 'woke mind virus'.[12]

Is it fair to label all three of them as transhumanists? All three entrepreneurs are public figures, who are highly conscious of their personal brands, and the term 'transhumanism' has received a rather dubious reputation in the public discourse. The 'World Transhumanist Association' even rebranded itself as 'Humanity Plus' in 2008. Neither in academia, nor in the public discourse, is the concept 'transhumanism' associated with positive qualities and traits. Dehumanization, inequality, eugenics, anti-democracy, oligarchy, totalitarian total control societies, and a necessary tension with religious, humanist, and ecological worldviews are often mentioned as reasons for why transhumanism has got such a bad reputation.

At the same time, it needs to be noted that by means of their intellectual judgements, personal connections, as well as some of their financial donations, their intricate connection with transhumanism cannot be doubted. Concerning Musk, this connection is particularly noteworthy with respect to the 'Future of Humanity Institute'. Besides many other connections, he donated £1 million to the institute in 2015.[13] He also contributed a blurb to Bostrom's book *Superintelligence* and popularized Bostrom's simulation hypothesis. Concerning Thiel[14] and Buterin, their connection with transhumanism is particularly strong when it comes to their donations to the Methuselah Foundation: 'Ethereum co-founder Vitalik Buterin donated $2.4 million ETH to the SENS Research Foundation, an outgrowth from Methuselah Foundation. Last year, he donated $25 million in SHIB tokens to the Future of Life Institute and 1,000 ETH and 430 billion ELON tokens to Methuselah Foundation.'[15]

Conclusion

These reflections have clearly shown that there is a great variety of different types of transhumanism, and it is essential to be aware of the various different approaches. Among the various versions of philosophical transhumanisms, foundational, original, classic, and Euro-Transhumanism need to be considered. Concerning entrepreneurial transhumanisms, Musk's version which is focused on existential risks needs to be highlighted in the same way as a version dedicated to increasing the healthspan which is supported by Thiel and Buterin. However, it needs to be noted that it does not make sense to apply philosophical criteria to entrepreneurial versions of transhumanism. At the same time, the current cultural, social, and economic relevance of transhumanism needs to be stressed. This applies to the realms of philosophy, culture, art, politics, as well as entrepreneurship. It is also intriguing to note that neither the entrepreneurs who affirm these ideas, nor the institution 'Humanity Plus' strongly highlight the word 'transhumanism', as branding has always been central when it comes to advertisements, and, unfortunately, transhumanism has closely been associated with many problematic terms so far. I think that the identification of classic transhumanism with the public perception of transhumanism is at least partly responsible for this situation. I share many worries concerning classic transhumanism. However, I also think that the fundamental principles of transhumanism increase the likelihood of a higher quality of life, which is the reason why I defend transhumanism, proposing a variant which I refer to as 'Euro-Transhumanism'.

Brave New Human

The universe came about 14 billion years ago and the earth about 4.5 billion years ago.[1] Then it took another billion years for the first life forms to evolve. Only in the most recent 530 million years have mammals, amphibians, reptiles, fish, and birds evolved. During this period, life also nearly became extinct again. The big five events of mass extinction occurred 440, 372, 252, 201, and 66 million years ago, with as much as 95 per cent of all marine life becoming extinct at times. All these occurred even before the first human had emerged. The last common ancestors of humans and apes lived about 6 million years ago.

Homo sapiens probably emerged about 300,000 years ago. At that time, due to a gene mutation, language development started, so that the prerequisites for language were present, which in each individual case then had to be realized again by means of a cultural and parental upgrade. Cultural control turns us into living beings who are actually capable of using language. With this upgrade, human beings came into existence. The language ability is therefore a parental upgrade, which is why Homo sapiens can also be called cyborgs (see Sorgner 2022a). The word cyborg stands for the cybernetic organism, the word cybernetic coming from the ancient Greek *kybernaetaes*, meaning the helmsman of a ship. Cyborgs are steered organisms, and humans are such steered organisms, which attained language ability due to the steering of a prior genetic mutation, and thus became rational beings. Each human being develops its own embodied reason due to an idiosyncratic control in each case (see Varela, Thompson and Rosch 2017).

Reason is not an immaterial spark but rather a result of a cultural upgrade. With the appropriate genetic mutation, we have become speech-enabled cyborgs, and as Homo sapiens we have always been speech-enabled cyborgs, reason being a part of the evolutionarily evolved body. The body, in turn, is made up of about 39 trillion non-human cells, most of which live in our gut, and about 30 trillion human cells. So, our body has more non-human cells than human cells. Even in other ways, the boundaries of our bodies are not so easy to determine. Our smartphones, for example, can be understood as an extended mind. A person with a cochlear implant can control it with his own smartphone and thus influence his own cognitive capacities. If the smartphone is taken away from him against his will, then this can also be understood as a violation of bodily integrity, since his cognitive possibilities are limited in this way. Even if one's own avatar in the metaverse is unintentionally harassed and touched by another avatar, then this can be understood as a violation of bodily integrity. The body, which is constantly in the process of change and in which reason is also present as a part, consists of more non-human than human cells and can possibly include silicon-based smartphones and digital avatars.

It is this image of Homo sapiens that is the dominant one, especially among critical posthumanists, but also among many transhumanists, in particular in Euro-Transhumanism. It is their central concern to discard the culturally created and prevailing image of humanity in our cultural sphere, which is based on the categorical ontological duality of an immaterial reason and a material body, since central structures of discrimination go hand in hand with it. Humanism based on categorical ontological dualities was culturally created in order to establish certain structures of domination. Central developmental steps of this cultural process are the anthropologies of Plato, the Stoics, Descartes, and also Kant. Important problematic processes of discrimination can be aptly explained by means of Plato's allegory of the cave (Sorgner 2022b). The bright sun stands for the unconditionally good, whereas the dark shadows far from it are identified with evil. This is how racism emerged as a cultural structure. The sun represents reason, which has been widely identified with the masculine, whereas unreasonable judgements are often based on typically feminine feelings. Thus, sexism has arisen structurally. Reason continues to be a quality that traditionally characterized humans. It does not belong to all other living beings. This is how speciesism came into being. Furthermore, reason enables one to recognize the unconditional truth in correspondence to reality. Thus, aletheism came into being, the discrimination of all ignorant people who think, live, and act far away from the only actual truth. One of these truths is that we should act according to our nature, which due to its essentialist demands, has discriminatory implications, too.

The central concern of critical posthumanists is to discard these discriminatory processes, which are based on encrusted cultural structures of a particular form of humanism. This is accompanied by numerous challenges that are sometimes not so easy to resolve. It is this approach that seems to me plausible, but still involves innumerable intellectual challenges that are not necessarily easy to deal with. There are also proponents of critical posthumanism whose objectives are highly problematic, for example, the radical posthumanists, who assume that global warming is the central contemporary challenge facing humanity and consequently argue as follows: The decisive factor for global warming is overpopulation. This gives rise to the moral and possibly also legal obligation to refrain from procreation or to regulate it politically. Such paternalistic and eugenic lines of thought are morally highly problematic and are in conflict with the fantastic cultural achievement of freedom. Those who see overpopulation as the central contemporary challenge must arrive at demands that conflict with freedom. There are other more promising approaches to dealing with global warming that are consistent with freedom and that involve technological innovation. Seeing technical innovations as the crucial tool to increase the personal quality of life is advocated by transhumanists.

In terms of intellectual history, transhumanism can be clearly distinguished from critical posthumanism (see Ranisch and Sorgner 2014). I share the central elements of transhumanism (see More and Vita-More 2013). However, radical variations of transhumanism, which are strongly highlighted in classic transhumanism, such as the concept of mind uploading, seem to me highly implausible and, more importantly, distant from what is socially relevant for us in the foreseeable future. The central transhumanist idea is that we should use technological innovations to significantly extend our current personal limitations, thereby increasing the likelihood of a flourishing life. Most transhumanists affirm a variant of naturalistic anthropology. Crucial to their self-understanding, however, are normative demands suggested by an evolutionary worldview.

Vaccines, antibiotics, and anaesthetics have been developed in the past 250 years. Public use of the internet has been around since 1990 only. The first smartphone was introduced in 2007. Today, it is already impossible to even imagine life without the digital world. The gene scissor, CRISPR/Cas 9, was realized in 2012. Musk's Neuralink, which researches brain–computer interfaces, was founded in 2016. The interaction of genetic engineering, brain–computer interfaces, and digitization is the promising area for transcending current personal limitations. Transhumanists call for a proactive grounding to promote these techniques in particular.

In Germany, transhumanism has become known through Sloterdijk's speech 'Rules for the Human Zoo' (Sloterdijk 2001) – who, incidentally, is not a transhumanist but a bioconservative thinker – and Habermas' implicit replica in his essay on liberal eugenics (see Habermas 2018). However, more than 25 years have passed since then, and transhumanism has evolved from a niche cultural movement of a small group of intellectuals to a widespread cultural phenomenon that has become indispensable in Silicon Valley, in Netflix series, in computer games, in Hollywood movies, US politics, and even in our everyday life.

Homo sapiens emerged 300,000 years ago. Will there still be humans in 300,000 years? If we consider the speed of recent developments, this is probably not the case. Either humans will be extinct by then or they will have evolved. In the past, it depended on natural selection who survived. Now we have the possibility of human selection. We can improve, we can enhance evolution. Should we do this? Since there is no moral merit in playing genetic roulette, proactive interventions are indeed the morally appropriate response to the newly given technological possibilities. There is debate among transhumanists about whether this is a moral duty, as argued by many utilitarian transhumanists, whether it should be an obvious choice, or whether it would be a vice not to resort to these possibilities. This spectrum of possibilities illustrates that transhumanism is far from being a homogeneous group of libertarian utilitarian hyper-humanists, all aiming only for going faster, further, and higher, to be able to produce a Superman on Viagra or a Wonder Woman with Botox. Another common representation among critics of transhumanism is to characterize transhumanism as a cultural movement aimed at uploading our personalities onto a hard drive to become immortal. Both of these false widespread representations are distorting caricatures that have nothing to do with a critically informed analysis of transhumanism. However, there are other prejudices regarding transhumanism that are an obstacle to a broader academic reception of this intellectual understanding of the world.

It is always implicitly conveyed that a critically thinking intellectual must necessarily be opposed to transhumanism. A small selection of inapplicable prejudices shall be briefly presented. In this way further facets of the 'new human' can emerge. The following four accusations will be briefly discussed here: (1) the totalitarianism accusation, (2) the vampire capitalism accusation, (3) the anthropocentrism accusation, and (4) the injustice accusation.

The totalitarianism accusation

When Habermas, in his essay on liberal eugenics, identifies transhumanism with Nietzschean breeding fantasies (see Habermas 2018), which he opposes with a pluralistic society based on a constitution, this very accusation of totalitarianism is hidden in his reflections. When in 2014 he puts transhumanist work in the context of sect-like thinking, he even strengthens this accusation (see Habermas et al 2014a). Contrast this with the reception of transhumanism among Russia's current leading intellectuals, such as Alexander Dugin, whose daughter was killed in an attack in 2022, and who is a mastermind of the truth of Eurasian Orthodoxy. Dugin explicitly calls transhumanism an idea of the devil. Pluralism and all the other evils of the world are manifestations of transhumanism, which is presented as the evil antagonist of orthodox truth (see Dugin 2021).

Habermas criticizes transhumanism as an anti-pluralist movement (see Habermas 2018). Dugin, on the other hand, criticizes it for its pluralism. Habermas' assessments are highly implausible. Transhumanism knows no rites, prayers, saints, or commandments. There are not even necessary dogmas. The opposite is the case. The focus is on scientific insights. When new results are available, old decisions must be revised and new strategies adopted. It is a pragmatic, liberal, and scientific attitude. Transhumanism is also unsuspicious at the political level. Politically, all leading transhumanists can be classified in the liberal spectrum, although the range is enormous. It ranges from libertarian orientations to social liberal ideals.[2] This does not mean that libertarian political orders do not run the risk of implicitly promoting paternalistic tendencies. This concern is certainly true. It is not the case, then, that a libertarian political orientation does not bring with it certain challenges. However, this fact is more rooted in libertarianism than in transhumanism.

Dugin's characterization of transhumanism, on the other hand, is already much more accurate as a description: pluralism, and the use of genetic engineering to promote personhood, are certainly norms that are widely shared within transhumanism (see Sorgner 2018a). The problem lies in his characterization of transhumanism as the devil's idea. His description of transhumanism is not implausible, only his normative evaluation needs to be turned upside down. It is precisely the anti-authoritarian, anti-paternalistic, and anti-totalitarian basic attitudes that are characteristic of transhumanism and that speak precisely for transhumanism. Dugin's evaluation of transhumanism as the devil's idea is the problem, less his description of transhumanism.

The vampire capitalism accusation

Especially from the left-wing political spectrum, one often hears that transhumanism is accompanied by capitalism without consideration of the next generations and sustainability (see Thomas 2024). It is certainly true that global warming is one of the important challenges of our day. However, is it the most important challenge to which all other considerations must be subordinated or not? This is the crucial question. If this should be the case, then eugenic regulations could be justified, the prohibition of reproduction, of flights as well as also of meat consumption. Of course, such decisions could only be made within a country, since, fortunately, there is no world government that can make global legal regulations. If a country bans coal-fired power plants and nuclear power plants, then such power plants will be built in other countries. The electricity or energy generated in this way is then still imported into the country where the bans were imposed. An effective solution looks different. Another approach is to tax non-sustainable operations more heavily. However, the same challenge as the one just mentioned applies to this approach. Moreover, it still promotes social injustice within a country. Meat, flights, and heated *fincas* only for the rich. Furthermore, it is important to realize that exponential population growth is the main reason for our enormous consumption and unsustainable actions, which are ultimately the decisive reasons for global warming. What are the options to address this challenge? A ban on procreation is a totalitarian eugenic measure that undermines the basic structures of a free society. Fiscal management leads to social injustice. Children only for the rich. Both measures are still not global approaches. The populations of China and India already represent more than a third of the world's population. Furthermore, it is important to note that with increasing wealth comes increased meat consumption. Eating meat is a sign of social distinction. Could (moral) appeals be an approach to dissuade people from meat consumption? A comprehensive, effective, and rapid change in action is unlikely to be achieved through appeals. What is needed, then, is an approach that works globally without the need for totalitarian regulations at the global level, which is neither desirable nor realistic.

University of Oxford economist and statistician Max Roser, who is also responsible for the outstanding website ourworldindata.org,[3] has shown that population growth occurs in four stages. With more qualified medical care and improved hygiene comes lower infant mortality and longer life spans. With increased education, improved quality of life and secure housing, food, and social security, the reproduction rate decreases. In Germany it is 1.5,[4] in Austria 1.4,[5] and in Italy only 1.2.[6] This phenomenon can also be observed globally. Studies by the United Nations even indicate that if the quality of life, hygiene, and access to education, healthcare, and basic food continue to increase as they have in the past, then it can be assumed that the 12 billionth person will never be born.[7] If this research is accurate, it means that overpopulation will never be a problem because innovations can be used to improve the quality of life in a way that significantly decreases the reproductive rate.

If these assessments are plausible, then globally no totalitarian, paternalistic, eugenic, socially unjust, or even inefficient measures are necessary to address the challenge of 'population development' and thus also the issue of global warming. Rather, the measures outlined here seem to illustrate that by means of innovation, the achievements which go along with freedom can be preserved and the challenge of 'global warming' can still be effectively addressed. This is not to say that other measures are not necessary. These considerations at least illustrate that we need not despair. There are studies that suggest that by increasing the quality of life worldwide, thereby also addressing the challenge of 'global justice', the interconnected challenges of overpopulation and global warming can be successfully addressed.

The anthropocentrism reproach

Transhumanism is only about increasing the quality of human life. It is a hyper-humanism. Transhumanism revolves solely around humans, even if it is spoken of overcoming of humans. These are also widespread criticisms.

They can be easily refuted concerning most versions of transhumanism, not necessarily some versions of classic transhumanism. A central concern of most transhumanists, for example many members of the Institute for Ethics and Emerging Technologies, is that personhood be extended to non-human animals.[8] Anti-speciesism and anti-anthropocentrism are central aspects of most transhumanist anthropologies. This thrust is not without moral challenges, yet it has practical implications. Merely, a few transhumanists hold anthropocentric ethical views, for example Steve Fuller (2011; 2012).

Carbon dioxide emissions from factory farming are a key factor in global warming. With an increase in the quality of life and the population rate in India and China, meat consumption there continues to rise enormously. Alternatives are needed. In Singapore, the sale of artificial meat was legally permitted not so long ago. Animals no longer have to be slaughtered. Carbon dioxide emissions from animal factories no longer exist. Soil contamination no longer exists. Antibiotics for the prevention of diseases no longer have to be given, which in turn has consequences for the risk of the coming about of antibiotic-resistant germs. At the same time, US researchers have developed vegan cheese made from real lactose.[9] However, the lactose does not come from cows but from genetically modified yeast. By means of genome editing, vegan milk can be realized, which is the basis for this cheese. These and numerous other examples can be used to show how technical innovations can be used to avoid anthropocentrism in action without us having to give up meat and cheese. Neither bans, taxation, nor advertising measures are necessary to change lifestyles but only better alternatives. In the case of these, the question of how they can be realized globally does not arise. Rather, the demand for such products increases automatically. It does not have to be forced. This is the great advantage of the concern to promote non-anthropocentric thinking by means of technical innovations.

The injustice charge

Subverting the principle of equality and promoting injustice were two major concerns of Fukuyama due to which he mentioned transhumanism when asked which idea he considered the most dangerous in the world at the beginning of this millennium (see Fukuyama 2004). An example will be used to highlight that the current handling of digital data in Europe is not in the interest of the people and that we need to radically rethink the meaning of digital data. This idea will only be briefly touched upon here. I have explained it in detail in my monograph *We Have Always Been Cyborgs* (Sorgner 2022a).

That data are the new oil is widely held by economists. Digital data are intellectual property. Oil, on the other hand, is a natural product. These are different entities. The assessment is nevertheless accurate, as both entities involve money, power, and influence. However, with the European General Data Protection Regulation, it is enormously difficult to collect digital data in a comprehensive manner in Europe. However, the relevance of digital data can hardly be, as they are needed for scientific research, technical innovation, the use of resources and energies, political decision-making, and numerous other reasons. In the US, digital data are collected by the big tech companies, and in China, in an even more efficient way, by the state, as data are also the basis for their social credit system.

The relevance of collecting digital data can hardly be underestimated, and Europe refrains from collecting, analysing, and using it efficiently. It is this basic attitude that is currently the central challenge for Europe. If digital data are as relevant as widely held by economists, then the General Data Protection Regulation will have significant consequences for Europe, leading to further economic decline, as innovation and research can only be efficiently realized in China. The necessary digital data are available there. If Europe is worse off economically, the middle class will probably be the first to feel it strongly. They will be either absolutely or relatively worse off than before. When this is the case, those affected look for culprits, which they usually find in the dissenters, the foreigners, the refugees, other minorities, or the evil other in general. The tension becomes more and more intense until civil war will break out. These are the prospects when financial resources are no longer available as they have been. It is the present legislation that will lead to discord, discontent, and turmoil in Europe. For this reason, a rethinking of the importance of digital data is essential.

My proposal is that personalized digital data must be collected by the state. Depersonalized versions of it can then be sold to companies, which can then analyse them. The analysis of personalized digital data should be done primarily by algorithms, since humans are easily corrupted. The key takeaway, however, is that the financial gain realized by the state through the algorithms should be used in the general interest. The interest that is widely shared is to live healthily longer, that is, a prolonged healthspan (Donner et al 2016, 353). The profits generated should be used, for example, to pay for public universal health insurance. Then digital data would actually be used such that it is in the people's interest. Currently, what is happening is that full consideration of digital data is being dispensed with in Europe, it is being used in the US to increase the wealth of tech entrepreneurs, and in China it exists to consolidate the authoritarian political order. Nowhere are digital data being used in the interest of the people. For this reason, the meaning of digital data needs to be rethought. It was important for me to mention this point briefly, because this issue is, in my view, one of the most relevant at the moment, on which it is urgent to reposition ourselves.

Conclusion

In this overview, my aim has been to highlight some of the central challenges to which posthuman thought draws attention. Along with the posthuman paradigm shift comes the need to discard numerous encrusted structures of dualistic humanism that have shaped Western thought and action over 2500 years. I look forward to further intellectual exchanges on these highly charged issues. Only through an open, critical, and pluralistic discourse is it possible to arrive at an appropriate consideration of the plurality of interests (see Sorgner 2019a).

Generative AI, Digital Data, and Superintelligence

Classic transhumanists are fundamentally concerned with a takeover by superintelligence.[1] Everyone has been talking about artificial intelligence (AI) at least since the release of ChatGPT. Hence, this topic needs to be dealt with by clarifying some fundamental issues.

AI is found in software that instructs a computer how to function, mimicking aspects of human intelligence. An algorithm is a type of software. The first one was written by Ada Lovelace, Lord Byron's daughter, in the 19th century for the planned Analytical Engine. An algorithm is used to solve a class of specific problems by means of a finite sequence of well-defined instructions.

If AI mimics human intelligence, to grasp the meaning of AI, we must clarify and define human intelligence, which is rather challenging. Human intelligence can be measured by intelligence tests. Maybe, intelligence is simply whatever gets measured in such a test. There are several different IQ tests. The Stanford–Binet Intelligence Scale has existed in standardized form since 1916. IQ tests typically investigate verbal, numerical, and spatial thinking capacities. However, it is unclear which thinking capacities should be included in such a test. Is there such a thing as emotional intelligence, musical intelligence, or creative intelligence? Some tennis players are said to play intelligently. Can we speak of a 'strategic intelligence' belonging to certain entrepreneurs?

If AI imitates human intelligence, it is important to have a comprehensive idea of human intelligence. This is even more important when distinguishing between various types of AI. It is common to distinguish between weak or narrow AI, strong AI or artificial general intelligence (=AGI), as well as superintelligence, which is a concept which can be traced to the British mathematician I.J. Good, who discussed the possibility of an intelligence explosion as early as 1965.[2] If an AI is embodied, it can access its environment through sensors.

We refer to 'weak AI' when we speak of AI capable of performing one specific task. An algorithm has beaten outstanding human chess masters since 1996, it has beaten humans in Jeopardy since 2011, and defeated a world-leading Go-player in 2016. These victories are impressive; however, it must be stressed that in all these cases, we are confronted with various versions of weak forms of AI.

Strong AI is a theoretical concept, possessing human-like capacities in problem solving, learning, or planning for the future. It is unclear what capacities an AI would have to possess to qualify as a strong AI. Would it only need to have human-like capacities applied to numerical, verbal, and spatial thinking, or would it also need to possess emotional, musical, or athletic intelligence? If emotional intelligence is a prerequisite to possessing human capacities, does it need to possess self-consciousness? If it needs to possess forms of musical and athletic intelligence, would it need to be able to play the violin or master tennis?

Superintelligence refers to even more developed capacities. It is defined as an agent possessing intelligence far surpassing that of human beings. It is likely that an intelligence explosion would need to occur to bring about such agents. It might also be the case that a technological singularity would precede the realization of a superintelligent agent. What would this mean in practice? Would superintelligent AI have to be better at playing the piano than Trifonov, better at playing tennis than Djokovic, and better at grasping human emotions than Dostoevsky?

So far, neither AGI nor a superintelligent AI exist. It is unclear what it would need to be capable of. The central issue here is whether it must possess self-consciousness, and whether it must possess certain physical capacities, such as the ability to play tennis or piano. It is difficult to speculate about this without actually having encountered such entities. This is where science fiction becomes relevant, as all these topics are discussed and dealt with intensely in science fiction novels, films, and television shows, particularly *Star Trek*, which is an incredibly valuable source.

AI and digital consciousness, and life

Two fictive *Star Trek: The Next Generation* universe characters leap to mind when thinking of digitization: (1) Data and (2) the Borg. Data is a sentient, self-aware male android, who serves as an officer of the starship *Enterprise* commanded by Captain Jean-Luc Picard. The Borg, on the other hand, are cybernetic organisms connected in a hive mind entitled the 'Collective'. These fictional entities are antagonists in the Star Trek universe. The Starship *Enterprise* belongs to the United Federation of Planets, whereas the Borg regularly show up as an enemy of the Starfleet. Both figures raise central philosophical issues. In the episode the 'Measure of Man', the show examines with philosophical nuance whether Data should be considered a person or property (Sorgner 2019c). Since Data is a robot, this issue is crucial to dealing with embodied AIs, in this case, a male android – whatever the description 'male' might mean in this context. Much more could and should be said on this issue, too. In this context, it is relevant that the creators of *Star Trek* believe that a self-conscious android can exist.

In *Star Trek*, Data was created by Dr Noonien Soong. Just as Data is brought into being, the Borgs are also products of invention. Being cyborgs, they cannot reproduce biologically and consist, therefore, both of biological and mechanical parts. Borgs have a group mind called the 'Collective' attached to a group consciousness that enables them to have group thoughts. Achieving perfection is their ultimate ambition. The organic parts of their bodies include microscopic machines – what we now call nanobots. They can alter someone's bodily chemistry or modify and repair someone's DNA. Borgs integrate other beings by means of assimilation, including both the injection of nanobots, called nanoprobes, as well as by means of surgeries to add mechanical implants to the biological parts. Besides assimilation, compilation might be another way for Borgs to come about. This might have been highlighted in the episode 'Q, Who', where maturation chambers, also known as Borg nurseries, are depicted. It leaves open the possibility of Borg also coming about by means of artificial reproduction and an artificial womb. As a consequence of assimilation, the process of artificial reproduction and related processes, the relevance of race and gender vanishes. The episode 'I, Borg' raises particularly further fascinating philosophical challenges which should be dealt with in detail. For our purpose here, what is central is the issue of consciousness. The Borg possess a group consciousness, as they are inter-connected in a hive mind through mechanical devices.

Both Data and Borg show that it is possible for mechanical parts to possess consciousness. This refers to a type of digital consciousness, existing on a silicon basis. This is the challenging issue that needs to be further considered. A significant amount of philosophical speculation has been devoted to the notion of 'silicon-based' consciousness in recent years, also inspiring many science fictional accounts of such as consciousness. It has also brought about many doomsday scenarios among natural scientists, entrepreneurs, and, most recently, computer scientists.

In 2015, an android named after a robot from a Philip K. Dick novel pronounced 'I'll keep you safe in my people zoo'.[3]

In 2014, Stephen Hawking told the BBC that: 'The development of full artificial intelligence could spell the end of the human race.'[4]

Elon Musk warned that AI could end the human race in 2024: 'I think there's some chance that it will end humanity. I probably agree with Geoff Hinton that it's about 10% or 20% or something like that.'[5]

The creators of *Star Trek* as well as these natural scientists, entrepreneurs, and computer scientists assume the coming about of a digital self-consciousness, algorithmic intentionality, and mechanical emotions rooted in silicon-based entities. However, whether this is the case is a challenging issue, deserving further intellectual reflection. I do not suggest that these qualities are not a possibility. I merely wish to highlight a few philosophical challenges that present themselves when we take digital intentionality, self-consciousness, and emotions for granted.

So far, self-consciousness, intentionality, and emotions can only be found in carbon-based entities. They do not exist in silicon-based entities. This means that it is unclear whether qualities existing in carbon-based entities can also become a feature of silicon-based entities. There are other grounds for supposing this to be the case. Maybe the strongest argument in favour of this assumption is the following:

Intentionality, self-consciousness, and emotions can emerge from inorganic entities. Big history reveals that the earth came into existence about 4.5 billion years ago.[6] It then took almost a billion years for the first forms of life to come into existence. What then happened was indeed a quasi-magical event. Only inorganic entities existed. Suddenly, due to an interaction between water, heat, and lightning, initial forms of life came about. Life differs from the inorganic insofar as living entities are capable of self-movement and obtain energy from their metabolisms. This means that they can eat, drink, and digest, which provides them with energy for existing. It is an incredibly intriguing process, and it is utterly unclear how life can have come about from inorganic entities. However, it has happened that due to special circumstances inorganic stuff was altered such that organic entities were brought about. These organic entities developed further into plants which can move towards the sun, magpies who can recognize themselves in a mirror, and human beings who are capable of reflecting on the origin of life. It has taken evolution about 3.5 billion years to develop from initial forms of life to the life forms surrounding us today. However, these processes show that it is possible for self-consciousness, intentionality, and emotions to emerge from an inorganic origin. If this occurred in the past, it could plausibly happen again in the future. So far, all life forms have a carbon basis. However, this does not mean that life and all other formerly mentioned capacities can only exist on a carbon basis. A silicon basis might be another possible foundation for such qualities (Sorgner 2020). There are already some indications that such capacities can be realized on a silicon basis. At least, Stephen Hawking mentions the following: '"I think computer viruses should count as life," Professor Hawking told the computer trade show in Boston.'[7]

Currently, neither regular viruses nor computer viruses are considered to be alive. They are capable of self-movement, but they do not possess a metabolism. They thus require a host on whom they depend to generate energy.[8] However, viruses seem to possess some qualities that are widely associated with life. Hence, Hawkings' suggestion that computer viruses should count as life forms had some reasonable basis.

Intentionality, self-consciousness, and emotions emerged from inorganic substances about 3.5 billion years ago. It is thus possible for qualities like these to develop from inorganic substances. In addition, one should note the existence of some silicon-based entities with life-like qualities, namely computer viruses. In addition, the plausibility of Moore's Law has widely been acknowledged, which states that the quantity of transistors in an integrated circuit doubles every two years, thus leading to an exponential growth and increase of computing capabilities. This is more of a self-fulfilling prophecy than a natural law. All these insights give justification for some intellectuals, like Ray Kurzweil, to suggest that the 'technological singularity' is near (Kurzweil 2005). It might be a prerequisite for the coming about of superintelligence.[9]

What should be noted, however, is that there is no proof of the existence of digital life forms. Even if they were to exist, it would require much more for them to develop from entities with digital consciousnesses to entities with digital intentionality, if such a thing is at all possible. However, for a sufficiently developed strong AI to direct an attack to wipe out humanity, it would need to possess intentionality. Hence, entities like the Borg with a conscious hive mind do not represent clear threats for us in the near future. It seems that we are still far away from such realities, if they can come about at all. The fervent discussions of Bostrom's book *Superintelligence* (2014), seem more a clever marketing strategy to promote the idea in popular media, or a means for acquiring third-party funding from visionary tech-entrepreneurs, than a serious philosophical attempt to deal with a pressing current issue. I think this book should be sold in the science fiction section rather than in the philosophy section.

However, both Data as well as Borg raise further philosophical issues that should be discussed in relation to the ethics of AI. I wish to focus particularly on the concept of the Borg, as their basic makeup is similar to research currently undertaken by Neuralink. Indeed, bearing in mind Neuralink's connection to enterprises such as Starlink and Space X, the problem becomes even more pressing. Dan Dinello identifies the Borg as 'contagious collectivist techno-totalitarian transhumanists' (2016, 83). However, he refers to the Borg's collective consciousness, which would only be legitimate in the case of the existence of a digital consciousness. Our reflections thus far demonstrate that this is not a pressing contemporary challenge. This does not imply that it is futile to reflect on the Borg, as the Borg indeed demonstrates significant similarities to central transhumanist enterprises.

The makers of *Star Trek* and many other thinkers, writers, and artists probed issues related to consciousness. However, there is no apparent reason why the organic parts of a cyborg Borg should suddenly no longer be primarily responsible for an individual consciousness, whereas the mechanical parts enable each individual to realize additional capacities and bring about a dominant collective consciousness. In addition, the nanobots promote the healthspan of a Borg, while other mechanical parts attached to the organism increase the likelihood of realizing additional capacities. Connecting the various mechanical parts to a Hive Mind seems to share some similarity with Musk's enterprise Neuralink, with the essential difference being that instead of being connected to a common consciousness, Neuralink enables a connection between the brain, the internet, hard drives, and algorithms. Instead of the hive mind determining the acts of an individual, Neuralink enables an individual to write, move, or play games. Of even more relevance, however, is that emotional states, psychological diseases, as well as cognitive activities can suddenly become accessible via the hard drive to which the brain is connected. Thus, depression, stress, and excitement can be correlated with acts, people, and situations. It is not the case that the individual vanishes, and a collective mind takes over when the brain gets connected to a computer, rather, the major concern relates to the privacy of feelings, as one's most intimate feelings can thus be absorbed and analysed by anyone with access to the corresponding digital data.

AI and the relevance of digital data

Musk's various projects are intriguingly interrelated. Unlike space programmes led by other leading tech-entrepreneurs, which offer the wealthy an intriguing space vacation, Starlink could turn Musk into a Bonafide Bond villain. The company is currently launching satellites into Earth's near orbit to collect visual data and analyse all parts of the earth and create a uniquely global space web provider. By offering a valid internet flat rate, Starlink could take over globally as an internet provider. Through Starlink, it might even be possible to use the same subscription on a transcontinental flight. Starlink enables users to have internet access at sea or in the most remote deserts. In the future, it may be one man who decides whether an army in a warzone has access to the internet: Elon Musk.

A second dimension must not be underestimated. By means of his satellite system, Starlink is able to permanently analyse the latest developments in real time. This applies to strategic military moves, weather phenomena, as well as the expansion of nuclear power plants. Elon Musk does not allow Ukrainians to use Starlink in the Crimea region.[10] Half of all active satellites in the near-earth orbits belong to Starlink already, and Musk continues to send up further small satellites.[11] This is an extremely clever move, as it increases his power in many different ways:

1. Starlink thus can become a global internet provider. By enabling internet access, digital data flows via his satellites, which gives him access to all the digital data collected. In the same way, the Chinese government can access

digital data of all companies active in China as well as some companies outside of China, for example TikTok, Huawei, or Alibaba, Starlink can access digital data generated by users from all parts of the world.
2. Starlink is a helpful spying device, as it enables the live monitoring of different kinds of visually accessible developments in all parts of the world.
3. Starlink is a means for generating new spatial property. According to Locke's property theory, one becomes owner of formerly unowned land if one works with this land and invests one's person into the unowned land. All territory on Earth is owned. Space is not infinite, in particular the space near earth, which is particularly precious due to being rather limited and useful. By being the first to send masses of satellites to the near-earth orbit, Musk has created new property, with lots of near-earth space becoming his property.

China has not failed to see the seriousness of these developments, and several Chinese military researchers have already stressed that Starlink satellites threatening China's national security would be shot down.[12] Starlink's potential for military use is enormous. Besides all the other uses already mentioned, it could simply increase internet communication between various air fighters with access to the Starlink system.[13] China has already demonstrated that it can destroy individual satellites. In 2007, they shot down one of their own satellites to send out a warning to other satellite owners.[14]

These developments show that the global war for digital data is moving to space, with near-earth satellite systems playing a central role in this conflict. These developments go further than this. Lonestar Data Holding Inc is ready to send devices to the moon to make research concerning the realization of lunar data centres.[15] This company has already raised $5 million in 2023 to create such centres.[16] Data can thus be stored in an area with relatively low temperatures, in centres far away from earth, enabling companies to generate new property. The fights to generate property in space are recent history. When discussing the future of digital data collection and storage, it is vital to discuss the moon as a data collection space, as well as the satellites bringing about the global internet, and brain-computer-interfaces like Neuralink.

The question of who has access to digital data, and who benefits from big data analysis and storage and collection of digital data is of significant philosophical import. We should not fear superintelligent entities putting us into a zoo, which is being highlighted by many classic transhumanists. We should also not fear a collective army of Borg enslaving or directing us through a harmonious collective hive mind. Yet, we should be concerned about the implications of access to digital data, which is a central issue for Euro-Transhumanism. We are already in a global war over digital data. Our future seems to be the collection of personalized, emotional, digital data globally being collected via Neuralink such that the information flows via Starlink satellites into a lunar data centre.

At the same time, brain–computer–interfaces such as Neuralink are also regularly discussed as an option for realizing a digital consciousness and the possibility of mind uploading. However, how is this supposed to happen? Silicon-based transhumanism usually embraces a functional theory of mind, which means that our mind is merely a function of our brain (Sorgner 2022a). This means that the mind and brain are related analogously to software and hardware, and in the same way as it is possible to install software on different hardware, the same holds for the mind. Science already tells us that the information stored in neurons can be passed on, as new neurons do permanently get generated anew, while it needs to be noted that 'neurogenesis is regionally restricted in the adult brain'.[17] However, neurogenesis exists, implying that information can be passed on to newly generated neurons. Whether this means that information only gets passed on to these new neurons or this process affects fundamental aspects of our consciousness is an open question. To determine whether mind uploading is a real possibility or not depends on the reply to this challenge. Yet even if it is logically an option, many questions are still far from clear, in particular concerning the issue about which functions need to be imitated by silicon-based computers to actually transfer consciousness.

Generative AI, innovation, and creativity

Digital evolution may be another means of possibly realizing a digital consciousness. In this context, the developments concerning generative AI are particularly striking. ChatGPT can permanently generate new texts in response to questions or commands. OpenAI's Sora can create videos up to one minute in length on the basis of a one liner as a prompt. What is needed both for ChatGPT as well as for Sora and other generative AIs is an immense amount of digital data, that is, texts in the case of ChatGPT, and videos, in the case of Sora. When a prompt is generated: 'create a painting of a white raven in the style of the later Picasso', the generative AI identifies central aesthetic features of the painting of the later Picasso and then combines them with central features of a white raven. Every time the same question is asked, a new image is generated. The same procedure applies if you ask Sora to create a video in which Beagle puppies play with an albino lion in the centre of Florence. It is also easy to generate a video in which Frank Sinatra sings Nirvana's 'Smells Like Teen Spirit'.[18]

To bring about such fascinating results, generative AI needs to draw on lots of digital data. What is at stake philosophically is that the data it needs to consider to realize such fascinating features can be both copyrighted material as well as open-access data. Here, the ethical challenge becomes obvious. In order to develop a new song in the style of Nirvana or generate a video of Taylor Swift dancing in the middle of a desert, it is necessary to draw on copyrighted material. This is highly problematic for artists, as it has taken designers and artists a long time, training, and preparation to develop the skills they have to enable them to create designs and artworks. Now generative AI simply draws on the data to generate songs, novels, or videos. Has AI developed the capacity to be creative? Is this an initial step for realizing a creative digital mind? Could such a development eventually bring about a conscious digital mind or is this already an encounter with the initial traces of consciousness in such generative AIs?

I think it is important to keep the achievements of generative AI in perspective. Although first encounters with ChatGPT are doubtless impressive, it has its limitations. Reading innovative thoughts by a leading scholar reveals refined nuances of which generative AI is not yet capable. This does not make generative AI's instrumental use any less fascinating, but a critical distance of an expert makes it more likely to distinguish human reflections from AI-generated ones.[19]

A central ethical question with regard to generative AI is whether it should have the right to draw upon copyrighted material or not. A reply to this challenge depends on an understanding of what generative AI does. Is generative AI learning when reading the novel *Elementary Particles* by Michel Houellebecq, is it incorporating the text such that it directly gets used by the algorithm, leading thus to copyright infringement, or is a different concept needed to appropriately describe what occurs? In order to become a famous writer, novelists read texts from Goethe, Dostoevsky, and Nabokov. Studying other artists, reading texts, engaging with artworks is not a copyright infringement. Any artist, writer, or designer needs to encounter these in order to develop their own capacities, to find their own style, and to realize what has happened in the history of art, music, and design. Reading a text, watching a video, and engaging with a work of art is a legitimate and important means of learning. Is generative AI learning when drawing upon a copyrighted novel by Houellebecq?

An alternative understanding is that generative AI is not learning but incorporating the material into their own algorithm. It is taking a copyrighted something and integrating it into its own structure to draw upon it for creating something new when asked for. The AI will use the shape of the McDonald's logo when asked to do so. It does not care what it draws upon but merely fulfils the function it is tasked with. However, is this a copyright infringement? An algorithm cannot become the holder of copyright.

Here, the decisive difference between human creativity and digital innovations can be discerned. Generative AI produces something new, whenever it is asked to do so. The outcome will be something new all the time. Generative AIs are clearly enormously innovative. However, there is a decisive difference between innovation and creativity. Innovation is simply the ability to bring about something new. Being creative has to do with an internal awareness of having created something special. Creativity goes along with a special state of consciousness, not just the ability to bring about something new. Creativity can occur only if generative AI realizes a new style and you realize its relevance. Then, you can take the innovative creation and work with it to realize a new creative entity. It might simply be sufficient to have consciously realized that what generative AI has produced is fascinating, intriguing, and relevant.[20] Then the human awareness of identifying the creation with something creatively special is the act of creativity. Generative AI merely brings about new forms all the time. Creativity is a human event that attends a special state of mind. If the creative event is directed at an artwork, then it furthermore depends on the recipients whether it is accepted as an artwork or not. If the art world accepts the creation as a work of art, then the work becomes a work of art. If this is not the case, then it simply remains the product of human creativity. Innovation, creativity, and reception are three categorically different processes that must be kept separate.

If this explanation of creativity is plausible, then the hope for bringing about a digital evolution such that a digital consciousness can occur by means of generative AI also seems to be rather limited. What we are confronted with concerning generative AIs are simply further types of weak AIs, which are capable of fulfilling one specific task. In the case of ChatGPT, the task has to do with the use of language. In the case of Sora, it has to do with the use of videos. Generative AIs are capable of innovation but not of creativity. What is missing again is self-consciousness, intentionality, the capacity of having feelings, desires, and longings, the will to survive, the will to power, and the will to create. I am not saying that an AI is incapable of developing these features, but so far, there is no clear indication that an AI will be capable of realizing such capacities soon. I am not a luddite, who doubts the relevance of emerging technologies. I am not a humanist, who upholds human ontological exceptionalism. However, classic transhumanists who permanently fear that AI will wipe out humanity are as far from a plausible understanding of ethical challenges related to AI as traditional dualistic humanists. AI intentionally killing human beings does not have to worry us at the moment. The issue is different concerning terrorists using AI for an attack. However, similar risks were already given with nuclear weapons.

Conclusion

In this section, I wanted to relativize the widely shared fears, worries, and warnings concerning AI. I hope it has become clearer by now that both traditional dualistic humanists as well as classic transhumanists do not present a plausible picture concerning the most relevant challenges when it comes to the ethics of AI. It is highly implausible that only human beings possess morally relevant capacities. Members of several non-human animal species clearly possess morally relevant capacities. It might even be possible that a sufficiently developed AI can develop morally relevant capacities. This is not the case so far. However, the widely shared fear among classic transhumanists that AI will gain digital consciousness, intentionality, and emotions soon, which will lead to AI attacking humanity to wipe it out, is not a pressing worry either. It is a great plot for science fiction movies, stories, or television series. It is a helpful marketing trick for getting talked about in popular journalism. However, it is not a currently pressing social challenge. The central ethical issues concerning AI have to do with digital data access, storage, and collection. Who has access to digital data, and for what does digital data get used? Neuralink, Starlink, and Space X may make Elon Musk a Bond-like villain. I have already explained why digital data are the new oil, as both have to do with power, money, and political influence. We have already seen the relevance of Elon Musk concerning the current wars. China, the US, and Elon Musk are the central players on this new playing field. When Musk highlights AI doomsday scenarios, he is distracting the wider society from the really pressing issues concerning AI, which have to do with many of Musk's enterprises. According to Euro-Transhumanism, we urgently need to rethink the meaning of digital data, rather than worry about a superintelligence that will wipe out humanity. It might be fun to think about a superintelligence, uploaded minds, and the simulation argument, but all these issues have absolutely no practical relevance for the central ethical challenges we need to face concerning AI.

The Case of Africa, China, and the Meaning of Digital Data

Here, I wish to mention one intriguing example concerning the relevance of digital data.[1] The wealth of the new age lies not in oilfields but in server farms. For quite some time, we have seen global fights for digital data. China seems to have the most promising strategies for gathering digital data. China created a digital firewall, which has destroyed the internet as a global entity. Chinese political leaders have the legal right to gather all digital information, and they use it for their social credit system. Outside of China, digital data can be collected by means of Huawei, Ali Baba, and TikTok. Digital data are relevant for financial, scientific, as well as technical flourishing. If bytes are barrels nowadays, then it is difficult to imagine how any nation can stop China from becoming the economically most flourishing nation worldwide. Economic flourishing also plays a role for political influence, which goes along with military power.

This analysis has various implications. The salaries go up in China, and products can no longer be produced as cheaply, as it used to be the case, when China was still extremely poor, only a few decades ago. However, they focused on innovation and the latest technologies and were able to realize an enormous economic growth. Now, cheaper workers, lower energy costs, as well as reduced expenses for logistical purposes are needed in Chinese companies. All these goals can be realized in various African countries. The salaries are cheaper, the sunlight can be used as an energy source, and the costs for transporting good to Europe are much lower than from China. In addition, many African countries have important natural resources to offer, too:

> Over a third of China's oil comes from Africa, as does 20% of the country's cotton. Africa has roughly half of the world's stock of manganese, an essential ingredient for steel production, and the Democratic Republic of the Congo on its own possesses half of the planet's cobalt. Africa also has significant amounts of coltan, which is needed for electronics, as well as half of the world's known supply of carbonatites, a rock formation that's the primary source of rare earths.[2]

It is a widely shared conception that the natural resources are the prime Chinese motivation when it comes to their interest in Africa, and again it is simply another story of colonialism and exploitation. However, the issue is much more complex. The following insight needs to be noted: 'the proportion of Chinese FDI[3] into mining (21 percent in 2020) is far lower than that of other countries such as U.K., France, and the U.S. (43 percent, 43 percent, and 37 percent in 2019 respectively)'.[4]

It is not the case that the natural resources do not play any role, but the other factors mentioned before, like cheaper work forces, play an at least as important role as the natural resources, if not a significantly more important one. To build and run companies, and to produce goods, what is also needed is a developed infrastructure and a high level of education. China is involved in many related projects: 'The central players in many of Africa's biggest ticket infrastructure projects – including the $12 billion Coastal Railway in Nigeria, the $4.5 billion Addis Ababa–Djibouti Railway, and the $11 billion megaport and economic zone at Bagamoyo – are being developed via Chinese partnerships.'[5]

Education is another field which does not remain untouched by China. For a long time, it has been the case that 'China also supported African countries that opted for communism and positioned itself as a global, moral role model'.[6] Yet, it also needs to be noted that 'since 2000 China's graduate scholarships and grants to Africans to study in its tertiary institutions have increased to about 61,000'.[7]

Over the same period, China has increasingly made its presence in Africa more visible by establishing more than 54 Confucius Institutes (CIs) and 27 Confucius Classrooms (CCs) across the African continent, according to the 2018 edition of *Quartz Africa*. Both CIs and CCs are major instruments designed to promote Mandarin and Chinese culture in Africa. 'The *South China Morning Post* reports that China has been highly successful in creating a cultural footprint across Africa, the world's fastest growing continent, through its Confucius Institutes.'[8]

Furthermore, recent studies also support the judgement that the aspect of cheap renewable energy is also relevant for the Chinese to invest in various African countries:

> One peculiarity of Africa's renewable-energy sector is the rapid increase and likely future growth of Chinese involvement in large-scale renewable-energy infrastructure projects. Insights from other infrastructure, utility and resource-extraction sectors in sub-Saharan Africa suggest that China is pursuing a specific Chinese model of investments characterised by enclave characteristics and including finance, turnkey project development and the importation of labour and equipment from China.[9]

All the previously mentioned insights reveal a double-edged sword. On the one hand, what gets imported is the Chinese culture, which implies hierarchical and paternalistic structures. The right of morphological freedom does not seem to receive any consideration within Chinese cultural structures. On the other hand, what seems to be promoted are a sustainable renewable energy sector, a functioning infrastructure which can bring about lasting changes, and most importantly an increased level of educational options. The realization of these changes might have a significantly stronger and lasting impact than the various development aid projects which Western countries have implemented in previous decades. If a complex infrastructure with a sustainable renewable energy sector can be realized, the foundation for a productive environment was realized. Due to the relevance of this sector for central global challenges like global warming, the options for exporting these technologies and further growth are enormous. Given the initial implementation of companies with the support of China, the possibility of local spin-offs can be realized, in particular if the educational sector gets promoted further such that human resources are available which can bring about such projects and put together successful spin-off companies.

A challenge which remains is that of a cultural and financial dependency from China. Various African countries have already reacted to specific attempts of creating such dependencies, for example:

> It all started in July 2020 when a Kenyan high court ordered the cancellation of a USD 3.2 billion contract between Kenya and China for the construction of the Standard Gauge Railway, saying the project was 'illegal' because the state-run Kenya railways failed to comply with the country's procurement laws.
>
> Also, Ghana last year scrapped a USD 236 million contract with Beijing Everyway Traffic and Lighting Tech company to develop an intelligent traffic management system, and the Democratic Republic of Congo is reviewing a USD 6 billion mining deal with Chinese investors, Reuters reported last month.[10]

At the same time, it must also be noted that a better infrastructure, a higher quality education, as well as functioning renewable energy sector would also go along with some social prerequisites which are of an enormous relevance for most human beings, for example, a well-functioning healthcare sector. Most human beings identify an increased healthspan either intrinsically or instrumentally with a better quality of life. It is actually the health sector in which there has already been a longstanding cooperation between China and various African countries:

> 'Health' is one of the oldest forms of engagement between China and Africa, dating back to the 1960s, when the first Chinese Medical Teams arrived on the continent to provide direct care to African

> people, build local capacity, train healthcare personnel, and construct hospitals, ultimately serving as a *tangible expression of solidarity among developing countries* against colonial and imperialist powers. Health cooperation, at that stage, fed into China's anti-imperialist discourse, helping Beijing to consolidate its network of regional partners. When the first Forum on China-Africa Cooperation took place in 2000, showcasing China-Africa relations to the rest of the world, 'health' appeared in both the FOCAC Declaration and Action Plan (the two official documents issued after each FOCAC). Though the focus on the health sector appeared *rather limited in content and relatively low in the priority list* (the last of a 10-point statement), its *soft power dimension* – i.e., China's long-held solidarity towards Africa – was *crucial in counterbalancing the rise of a new, perhaps less benevolent, logic of interaction based on economic interests*, and achieving the associated foreign policy objectives.[11]

This development has even been intensified in more recent times:

> Beijing managed to achieve a number of objectives during the Covid-19 pandemic, somehow reinforcing its health strategy. The distribution of vaccines exemplifies this: donations of Chinese vaccines to Africa are significantly lower than commercial sales (6.7 million donations versus 47.5 million vaccine purchases), but Africa remains the *second largest recipient of Chinese Covid-19 vaccine donations worldwide* (after the Asia Pacific region, which takes up to 65% of the total share), showing Beijing's commitment to help the continent. At the same time, out of the 33 million doses delivered to Africa thus far, half have gone to China's strategic allies, Morocco (16.5 million), followed by Zimbabwe (4.4), Egypt (4.1), and Algeria (1.8), with the delivery gaining momentum over the last few weeks.[12]

By mentioning the vaccinations, I merely stress the Chinese interest in being connected to the African countries mentioned, the entanglement between China and Morocco and some other countries. It should not be much of a surprise that China is interested in Morocco, given the geopolitical importance of this country. It is connected to the Mediterranean Sea, as well as the Atlantic Ocean. If goods get produced there, or they arrive there from other African countries, they can easily be shipped further either to one of the Mediterranean countries or the Netherlands, the UK, and the Scandinavian countries. The goods can be produced cheaper due to lower salaries, renewable energies, as well as cheaper logistical costs. It makes an enormous difference whether you have to transport tablets and smartphones from China to Europe or from Morocco.

From what we have realized in these reflections, there is a complex entanglement between China and various African countries which has been intensified significantly in most recent times. It is closely related to ethical, political, and cultural issues concerning emerging technologies. There are several serious challenges which go along with these developments, which have to do with paternalistic structures. These are serious issues. Yet, it depends upon the force, and relevance of these structures on how problematic they are. However, given the relevance of technological innovations for the economic, political, and entrepreneurial flourishing of a country, an enormous potential for sustainable changes have to be noticed, too. Given the enormous relevance of digital data, and the variety of industries which are related to the processing of digital data, all these developments might promise a flourishing future for the countries in which these innovations get implemented.

Conclusion

Digital data are power, and the countries which manage to collect the data most efficiently seem to have the most promising outlook for a flourishing future. Silicon-based transhumanists talk a lot about the implications of digitalization, whereby topics like mind uploading, a digital consciousness, as well as the simulation argument are discussed most often. All these topics are fun to talk about, and they can enable intellectuals to get a clearer understanding of who we are and how we relate to others. However, from a pragmatic perspective none of these issues is of any relevance. If one takes the issue that a sufficiently developed general AI will put humans into a zoo too seriously, it distracts us from being focused on the philosophical issues which are of an enormous pressing relevance. It seems to me that the following question is the most central one, and it is this issue which gets highlighted by Euro-Transhumanism: How can we collect and use digital data in a democratic manner, as digital data are of central relevance for scientific, technological, as well as political endeavours? It is already the case that Chinese scientists publish more peer-reviewed articles than US scholars.[13] This is no surprise, as data are needed for studies in social sciences, the natural sciences, as well as engineering. China can collect data within China as well as outside of China. US American companies cannot collect the data so easily, as they cannot simply legally force people to share their data with them. This is the reason why the financial, scientific, as well as political flourishing of China seems to be extremely likely in forthcoming times. Given the enormous interest of China in selected African countries, it seems probable that it will have some promising spill-over effects for these countries concerning education, infrastructure as well as employment. These developments also seem to be sustainable as they are related to some long-term decisions, changes, and investments. Will this be in the long-term interest of these countries? This issue is a separate challenge, which is related to the question which interest is most relevant? Is it freedom or financial well-being? How successfully will China manage to import their own cultural structures into the previously mentioned African countries? Concerning educations, infrastructure, sustainable energy, healthcare, employment and related issues, the Chinese investments seem to be in the interest of the countries mentioned. A question which still needs to be analysed is the following: What will be the implications concerning the political and cultural structures which get imported at the same time?

We Have Always Been Cyborgs

An essentialist human anthropology has been dominant in Western culture at least from Plato's times until Kant presented his ethical thoughts. It is still the case that an essentialist human anthropology is widely shared in all parts of the world. Since 1869, the Catholic church has upheld that fertilization goes along with animation which implies that the immaterial rational soul and body get united then. God is reason, and so the human image is with all human beings from fertilization onwards. Consequently, all human beings possess dignity from that time onwards, which implies that they ought to be morally respected. It was a strong concept of dignity, as it demanded universal validity (Sorgner 2010c).

An alternative understanding has come about with Darwin, Nietzsche, and Freud. A shift towards a non-essentialist anthropology has occurred with these thinkers. The world has become an incredibly dynamic one with these intellectuals. Six million years ago, the last ancestors of apes and human beings lived. A gene mutation occurred 300,000 years ago, which led to the coming about of human beings, as it enabled humans to develop the capacity to develop language. There is an intricate interconnection between reason and language. Hence, human beings did not receive reason from God at fertilization, but reason is the result of a genetic mutation which implies the psychophysiological capacity of developing a language, which again and again gets developed as a consequence of the appropriate cultural impact. This understanding implies that each human being has a unique reason, which, on the other hand, is sufficiently similar to all other reasons such that it enables us to communicate successfully, even though a comprehensive understanding in all nuances can never be realized.

Given a non-essentialist ontology of permanent becoming, there is no foundational reason why any entity deserves to be respected (Sorgner 2022b). Norms, values, and a moral status are the result of human interactions, and as consequence of a certain power dynamic normative fictions get established. We agree that certain entities deserve to be treated with respect and others do not. However, it is a weak foundation for personhood or dignity, as it no longer rests on a criterion which demands universal validity (Sorgner 2022a). This weakness or perspectival epistemology is actually a strength, as it demands a certain humility concerning one's own perspective, an increased respect of moral laws of other cultures, as they might be correct, and it avoids neocolonial tendencies within intercultural discourses (Sorgner 2007).

There is a permanent struggle during which redefinitions occur of what dignity and personhood should mean. A central thought in this context is who deserves to be respected. It seems that entities do not wish to suffer. Hence, suffering seems to be a morally relevant capacity. In order to accommodate this insight a certain group of human beings have agreed that entities should be granted personhood on the basis of their capacity of suffering. As there is a hierarchy of capacities of suffering, the question arises whether there should also be a moral hierarchy concerning personhood. Another question is how suffering can be compared in the appropriate manner. In addition, the issue has come up whether suffering is limited to carbon-based entities or whether it can also apply to silicon-based entities. What is needed for suffering? Consciousness, self-consciousness, a brain, a nervous system, or simply cognition? All these philosophical challenges become even more difficult given the anthropological revisions which have taken place.

Human reason is no longer attached to the body as an immaterial something, but it has come about on the basis of evolutionary processes. Thus, embodied reasons have been developed. Furthermore, our body consists of 39 trillion non-human cells but only of 30 trillion human cells.[1] Without the non-human cells, the ones which do not contain our DNA, human beings could not exist. Are they part of our body? What about our smartphones? Someone with a cochlear implant needs the smartphone to adapt her cognitive capacities. Is it part of that person's bodily integrity. An avatar can be intensely connected to someone's existence. If some other avatar touches it in an inappropriate manner, is that an infringement on someone's bodily integrity? Language has become a part of our psychophysiological make-up. Thus, we got upgraded towards language-enabled entities. Hence, we have always been cyborgs, as cyborgs consist of the words 'helmsman', ancient Greek *kybernaetes*, and organism. We have always been steered organisms, such that we have become language enabled entities.

Interfaces, actuators, intentionality, and agency

The issue is becoming even more difficult by further analysing the silicon–carbon interfaces. Based on a traditional understanding, the limits of the carbon-based body are the limits of someone's personhood, which needs to be respected. Someone can be removed from a ventilating machine, which supports the survival of a person, and it counts as passive euthanasia, as thus the natural death process can occur. However, if the ventilating machine was seen as a part of someone's personhood, disentangling a person from the ventilating machine would be comparable to removing someone's organ, which would count as murder or active euthanasia, depending on the circumstances. Here, the question arises under which circumstances a computer–body interface should count as a unified person, which needs to be respected. This issue also applies to the question concerning the status of cybernetic avatars.

AI algorithms are widely referred to as artificial agents. This also applies to embodied AI algorithms which have sensors as well as actuators. These entities are referred to as artificial agents when they act, which means that they can do something. Replying to a prompt and formulating a detailed response is one type of action. If an embodied AI moves from a to b, it also counts as an action. Here, we have the issue of translating human intentions into an AI action, as AI agents run based on information produced by human intentions. The question arises how such a translation can be realized and whether it can be realized appropriately, as analogous information needs to be transferred into digital information. There is always some kind of alteration when this occurs. In addition, it needs to be highlighted that artificial silicon-based agents are significantly different from carbon-based agents. Carbon-based agents get hungry, angry, or aggressive, which brings about their intentions to act. This is not the case with artificial agents. Their acts are based upon an impact from a carbon-based agent, which then needs to be translated for the actuator such that artificial agent can do something. The intention arises in a carbon-based agent, and their acts are triggered by an actuator which translates an intention into an action. In how far the translation process is successful is an open question. The traditional agent might have had an unclear intention. Due to a lack of sensorial impact, an insufficiently structured algorithm, or a low degree of data, the actuator might not actually bring about an act which fully corresponds to the intention of the non-artificial agent.

One challenging philosophical issue is whether these challenges arise only if analogue information needs to be transformed into digital information or whether this also applies within a carbon-based body; for example, if a human agent intends to move a toe, a muscle in the toe brings about decisive movement. Can a digital actuator be compared to a muscle in the body? Sometimes, you want to move your toe, but it does not work out. Sometimes, you lose control over your muscle, and it moves, even though you do not want it to move. Sometime, the sensors on the toe enable you to feel pain and sometimes this is not the case. These intuitive examples provide some reasons for claiming that the translation process between intended act and actuator does not have to be successful in the case of algorithms, and the same can also be the case in the case of carbon-based entities like muscles. An intention is needed for bringing about an action. However, neither a digital actuator nor a muscle can have an intention. The intention arises somewhere else. Here, the brain comes in.

The brain is a particularly interesting entity, in particular when reflecting upon the notion of a digital agent and artificial intelligence. A general sceptical response often comes up, when I mention that digital agents lack intentionality, consciousness, self-consciousness, the experience of pain, pleasure, and emotions, as well as the perception of colour. This response usually stresses that it is implausible to apply anthropocentric concepts to artificial entities, as a digital mind might be something entirely different. Point taken. However, this sceptical criticism can always be made, and it cannot be refuted. I cannot exclude the possibility that the glass in front of me is staring at me and is in pain. What is decisive, however, is that there is no indication of this being the case. Being a human, I can only apply my own criteria to concepts. I cannot move beyond the limitations of my own capacities. What does this imply?

It seems to be the case, and in medicine, it has widely been agreed upon that if the entire human brain becomes dysfunctional, then self-consciousness is irreversibly lost.[2] In principle it might be the case that by means of bioprinting the reasons for these dysfunctionalities can be made undone so that self-consciousness can be regained, but so far this has never been realized. Hence, I do not have an indication that this can actually be the case. Given the observations concerning whole-brain death, a new concept of death has been widely accepted. Consequently, it seems to be the case that consciousness and self-consciousness are somehow correlated with a certain complexity of the brain as well as the existence of a nervous system. All entities which demonstrate self-consciousness, for example by passing the mark test concerning the option of recognizing themselves in the mirror, have a sufficiently structured brain as well as a nervous system. Hence, the concept of self-consciousness and a certain mind seem to be correlated with a certain brain complexity, the existence of a nervous system as well as a carbon-based body. This does not mean that silicon-based entities cannot have a mind. However, so far there does not seem to be a silicon-based entity which shows capacities which are widely identified with mental capacities. An autonomous car does not simply decide to drive to the seaside as it enjoys the fresh breeze of salty air there.

This thought shows that even though a general sceptical observation cannot be refuted, there are clear indicators which go along with entities which have a mind. It might be the case that the table in front of me has a mind, too. However, it does not possess any of the indicators which usually go along with entities which seem to have a mind. Based on these Euro-Transhumanist reflections, it makes no sense so far to talk of artificial minds. Neither the actuator in an embodied AI, nor the muscle in a toe can have an intention, because an intention seems to arise solely in a carbon-based entity with a nervous system, and a sufficiently complex and functioning brain so far. However, neither an actuator, nor a muscle has to always do the action the carbon-based agent intended. There can always be translation issues. Given that this is a plausible description, it follows that there does not have to be a substantially categorical difference between an artificial agent with an actuator and a carbon-based actuator like a muscle. However, a proper agent needs intentionality and all the other capacities which go along with it.

Open- or closed-source algorithms

In any case, there are some further issues concerning the actuator and the processes which take place in it. It needs to possess an algorithm, which can be an open-source or a closed-sourced algorithm. The algorithm could be a Go algorithm, in which merely the fundamental rules for go are included. Then, it can undergo an artificial learning process by playing against another such minimal algorithm, again and again and again. Thus, it learns, and it gets filled with data. The more data it contains, the better it gets at playing Go, as it has had artificial experience in going through an incredibly big number of different possibilities before. Thus, it can calculate 50 moves ahead, which is a capacity human Go players do not possess and cannot possess so far due to their more limited mental capacities. Maybe they can calculate 20 moves ahead. Given an algorithm as well as a great amount of data, it is well-equipped to deal with new sensory inputs, and it can act in the best manner possible to fulfil its goal, namely, to win the game.

There are many challenges when it comes to the issue of actuators, in particular with respect to digital data and algorithms. One challenge concerns the topic of closed-sourced versus open-source algorithms. Most crucial is the question concerning oversight and embedded restrictions. In a closed-sourced algorithm, the owner decides upon and limits the collaborative potential, which hinders innovation and radical breakthroughs. However, coders have a great variety of different intentions. While, so far, we do not have to be afraid of artificial general intelligence (AGI) or superintelligence taking over, there are dangers related to the use of AI.

A brief side note on why we do not have to be afraid of AGI or superintelligence: AGI and superintelligence demand that an AI can imitate all types of human intelligence, maybe all elements of a human being. This seems to presuppose digital life, digital emotions, digital consciousness, and digital intentionality. However, none of these concepts is meaningful so far. Concerning human evolution, it took about 3.5 billion years for initial life forms to develop into highly complex self-conscious agents. So far, there are not even any forms of digital life. To regard an AGI as a central existential threat lacks any solid basis. Hence, seemingly philosophical books on superintelligence should be moved from the philosophy section into the science fiction section of bookstores. Superintelligence will not be an existential threat soon.

However, there are dangers related to AI, and they have to do with algorithms. Here, the issue of closed-sourced versus open-source algorithms can be relevant. ChatGPT is not open source. It is extremely user friendly and efficient. However, it is also helpful that the owners have introduced certain limitations into the algorithm. A detailed description on how to build a deadly airborne virus by means of substances which can be purchased in regular drugstores could have devastating consequences. Algorithms can become dangerous, depending on who is asking the prompts. It could provide enormously helpful information for terrorists, too, but then the dangers for humanity are enormous. This is an issue which is more pressing concerning open-source algorithms, as no one oversees their use, and no one has imposed limitations concerning which question may or may not be asked. Then again, the question arises, who should have access to the unlimited algorithm of a closed-sourced algorithm. Do you need a police clearances certificate? Should it be obligatory for users to be without criminal record? Do you need to have passed a psychological test which certifies that you have the appropriate qualities for such an enormous responsibility? The more people have the right to access them, the more dangerous the situation becomes. However, to creatively develop the algorithms further, coders need to have access to the unlimited version. Luckily, developing such dangerous algorithms on the basis of open-source algorithms goes along with many practical challenges, which are not so easy to solve as they have to do with technical skills as well as the availability of the appropriate digital data.

Actual and artificial digital data

Besides algorithms, the efficiency of an AI depends upon the digital memory it possesses or, in other words, the digital data on which it runs. The more data, the faster, and reliable the response of the algorithm. Bytes are barrels nowadays, and we are already in a global war on digital data. The system knows how likely particular moves are and chooses the move which leads to realizing the goal for which it was programmed. The question is from where to get the best possible digital data. The most efficient systems so far seem to be the Chinese version, which basically forces all companies which exist in China to share their digital data with the government. Given the size of China and all its inhabitants, given that Chinese companies like Alibaba, TikTok, or Huawei also function outside of China and collect data there, and given the enormous technical skills which companies in the digital sector possess, it seems likely that China has the most efficient system for collecting enormous amounts of digital data. The libertarian market model which is dominant in the US is doing well, too. As the US government is aware of the relevance of digital data, it decided to implement Stargate, a new company which plans to invest $500 billion in AI infrastructure.[3] The model of the European Union, on the other hand, seems to undermine all central European interests, as the GDPR, the general data protection regulation, makes a comprehensive collection of digital data enormously difficult – all in the name of privacy concerns. I am not saying that privacy concerns are unfounded, by the way. Whenever human beings get hold of data, which they can use against others for their own benefit, they will do so. This is the reason why human access to digital data needs to be minimized. However, this does not have to imply that digital data should not be collected, if it has been widely acknowledged that in the new economy, bytes are barrels. The question remains how this should occur. The authoritarian Chinese model is not in tune with the achievements of the Enlightenment. A libertarian model leads to a rigid hierarchization in society, such that the freedom of the less well-off gets limited. However, if freedom undermines itself, then negative freedom needs to be balanced, for example with other norms like equality and solidarity. Rather than permanently collecting more and more digital data, it has often been suggested recently to use AI-generated artificial data instead. However, this does not seem to be the most reliable way to deal with this issue.

AI-generated artificial data will produce an enormous number of possible versions of a melanoma, many of which might never occur in real life, and the ones which actually do show up in real life might not come up either. This is the challenge in question. Given these possibilities, a melanoma detection AI would be capable of detecting many versions of melanoma which will never actually come into existence but might not detect some which actually do exist. What health researchers are interested in, are the melanoma which actually do exist in real life, and they are interested in the likelihood of their existence.[4] Proper data collection is needed for their availability as digital data. In addition, to grasp a new disease, digital data concerning this specific disease is needed, and not artificial data of an enormous number of possible versions of diseases whereby the great majority never has or never will come into existence. However, the new disease which indeed kills people might not have been generated by artificial data. I am intrigued by the possibilities of artificial data; however, only a comprehensive collection of digital data allows us to have a grasp of which data are actually relevant for our lives. Artificial data might provide us with an enormous number of possible scenarios, but, thus, it does not have to take into consideration the scenarios with which we are actually confronted.

In the actuator, algorithms and digital data are two central features which need to be considered, in particular the relevance of collecting digital data can hardly be underestimated. Despite everything which has been said so far on the differences between human agency and artificial agency, it might still be possible that the relevance of intentionality, consciousness and emotions has been overstated.

What are human agents influenced by? The intentions of a human agent are based on genetic as well as environmental influences as well as a myriad of complex interactions between them. In a way, the same could be said of an artificial agent. The algorithm bears similarities to the genetic imprint, which then acts based on its material form. By means of acting, it generates digital memories aka digital data. The more experience is given in the form of data, the better the original task of the algorithm can be realized. Hence, an algorithm with less experience can work significantly differently to an algorithm with a lot of experience. The same applies to human beings. Experiences alter human decisions. Nevertheless, there seems to be a difference between human and artificial agency, as having self-consciousness seems to make a difference. By means of self-consciousness, carbon-based agents can adapt their goal, at least it might seem thus. A certain goal might not be as pleasant, as it seemed originally, and it might become clear that a completely different state might be something utterly fulfilling. If such a radical change occurs does the human algorithm get altered or is it still the same human algorithm, for example, human happiness or power, which gets realized by different means? It might merely be the case that due to our consciousness; a more complex algorithm is at work. This possibility is definitely an option which cannot be excluded. This additional level of complexity might still justify a significant difference between human and artificial agency. However, the final word has not yet been said on this issue, as the verbal as well as visual acts which generative AIs are already able to realize, are enormously impressive indeed. Nevertheless, unless I see an autonomous car deciding to drive to the coast to listen to the relaxing sound of the waves, I have reservations concerning a simple-minded identification of human and artificial agency.

Even given the powers of AI so far, the implications of ChatGPT and related generative AIs and the future of scholarship in the humanities as well as in the arts need to be considered, too. A simple-minded banning of the use of AI, or related regulations, cannot be the appropriate response to this new tool, as the advantages related to these new possibilities cannot be underestimated. In the same ways, as scientists started reading printed books, writing on digital typewriters, and researching papers on Google, the advantages of integrating AI in research need to be taken seriously, too.

It is interesting to note how the German composer Sven Helbig recently dealt with AI. Many AI projects in the past have tried to create music in the style of Bach or Mozart. There has been a fascinating Beethoven AI project led by Matthias Röder which used the available fragments for Beethoven's 10th symphony as well as AI to create two movements of the 10th symphony, which was an enormous success.[5] Sven Helbig then considered that instead of AI being applied to the works of a dead composer, this could also be the case for a living composer, namely himself, and he used AI for various musical pieces of his project entitled Skills.[6] He used AI to realize several versions of various of his musical pieces of this project, and in a few cases the outcome was extremely intriguing. It was not such that he was fully content with them, but the overall structure was fascinating. Hence, in some cases, he took the AI generated structure, and started working with it, such that he ended up adding and revising the AI-generated structure. In the end, the result was such that some of the musical pieces were so refined that he decided to use them instead of the original musical pieces which he had created by himself. It is this way of progressing which might also provide an alternative way of dealing with generative AI in the humanities, both concerning student papers as well as for research undertaken by researchers. What generative AI can do is to permanently produce innovative output. Some are more intriguing than others. However, it is excellent for brainstorming and experimentation, and users might get confronted with more innovative ideas than they could come up with by themselves. To understand the flaws of what AI has produced, you often need to already have certain intellectual capacities in the field. AI can definitely help during the brainstorming process. Yet, then the proper work comes, which needs to be done by the human researcher. It is a matter of selecting, highlighting, fine tuning, refining, and explaining things in more detail. Generative AI thus becomes an extremely useful device, which gets incorporated into the cyborg creator.

Critical text editions of the collected writing of an author have enabled readers to get a critically informed understanding of the writings by a thinker. Digital text editions have radically altered the way we can analyse texts. Data analysis tools have enabled us to get insights into the thinking of an author which would not have been possible in earlier times. Generative AI is simply a new step which again provides us with new possibilities. It is an extremely helpful device for pushing human creativity to a new dimension. Human creativity does not have to be understood as the process of bringing about new forms. This is what I call an innovation. Human creativity is the act of choosing one AI created outcome rather than one of the many others, as it is particularly valuable. Then, the human needs to take it, refine it, polish it, and work on all the details. I think this revised way of dealing with innovation and creativity will play an enormously important role in the future of the humanities as well as the arts.

Conclusion

In the previously discussed reflections, I have merely highlighted a few central challenges when it comes to the human–machine interaction. To realize that we have always been cyborgs is an epochal occurrence. The Western cultural tradition has been dominated by discriminatory, violent essentialist concepts of dignity and personhood. With Darwin, Nietzsche, and Freud, a paradigm shift has occurred, which has brought about the plausibility of embracing ethical non-essentialism. Thus, our norms and values have become weak, as they turn into contingent nodal points. They are interpretations which can be wrong, but which do not have to be wrong. It becomes relevant to realize that this weakness is an enormous strength, as it implies a certain humility concerning one's own perspective: I live on the basis of my own criteria. However, I need to consider that I might be wrong. This attitude also leads to an increased respect towards the moral laws of other cultures, as perspectivism implies that others might be correct. It becomes appropriate to enter a discourse. Last, but not least, the possibility of discourses between cultures gets promoted, as a self-distancing stance towards one's own values and norms avoids neocolonial tendencies within intercultural discourses. Thereby, a lot of violence in intercultural exchanges get avoided. However, the reflections on all these tricky issues have only been initialized recently.

Euro-Transhumanism: *Ex Positivo*

Transhumanism is a basic intellectual attitude that can be traced to Julian Huxley, who coined the term in 1951 (Huxley 1951).[1] 'Transumanar', a related verb, had already been developed by Dante in the *Divine Comedy*. Although the word 'transhumanism' predates Huxley, prior to him, it was associated more with the otherworldly developments of human beings, and its meaning changed after his coining of the term.[2] Transhumanism is a basic orientation towards the significance and use of technology. It is wrong to assume that transhumanism always goes hand in hand with mind uploading, that is, the transfer of the human personality to a hard disk (Sorgner 2020). Otherwise, Julian Huxley himself would not be a transhumanist. While this mind uploading is affirmed by many transhumanists (for example, Ray Kurzweil), it is by no means shared by all. Even more absurd is the image of all transhumanists striving to become immortal by means of mind uploading. Critics of transhumanism, both from a humanist as well as from a critical posthumanist perspective, present transhumanism in this way to highlight the absurdity of this basic stance. Such polemical statements are not informed by critical inquiry or analysis. This worldly immortality cannot even be thought of consistently. There are two likely developments for the universe: it will either cool down, which could be accompanied by a cosmic standstill, or there will be a cosmic singularity, that is, a point of enormous density, containing all the energy after a contraction of the universe. How could a personality uploaded to a hard drive survive such a state? It is not even internally consistent to conceptualize this-worldly immortality. Serious transhumanists would not uphold immortality in a literal sense. This does not mean that transhumanists do not use the word 'immortality' (Sorgner 2022a). When they do, they usually do so for rhetorical reasons, to draw attention in this way to the importance of extending the healthspan. It is this claim that is, in fact, advocated by most, if not all transhumanists. It should be emphasized here that it is not the lifespan but the healthspan that is spoken of, that is, the time during which persons are alive and healthy. This objective is affirmed because it is assumed that personal quality of life is promoted in this way, and promoting this is, after all, a primary transhumanist concern.[3] In general, the central concern of transhumanism can be formulated as follows:

> It is in our interest to push our current boundaries in a significant way using the latest techniques, so that the likelihood of the coming about of the posthuman can be increased, as this will increase the probability of individuals achieving a better quality of life.

The following example can be used to illustrate the plausibility of this basic attitude. Over the past 200 years, the human lifespan has almost doubled.[4] In this way, the quality of life has also been significantly increased. It should be highlighted that transhumanism's primary intention is not to increase the lifespan but to extend the healthspan. People associate a great plurality of goals with a fulfilled life.[5] For some, enjoying oneself is important, for others, it is establishing a dynasty, and still others identify a fulfilled life by leaving behind a cultural legacy. Identifying commonalities is difficult, which is why the category of morphological freedom is central to transhumanism (Sandberg 2013). However, it should be acknowledged that different transhumanists uphold different concepts of the good. Classic transhumanists affirm a hyper-renaissance concept of the good, for example, Bostrom (Sorgner 2022a). The concept that no non-formal concept of the good is plausible is characteristic of Euro-Transhumanism, as suggested in my own approach. Beyond these considerations, we must affirm the importance of health as a precondition for leading a fulfilled life. This statement does not need to hold for all. Most people identify health with a successful life in one way or another (Donner et al 2016, 353). For some, it is of intrinsic importance, so being healthy is directly correlated with a high quality of life. For others, health only plays an instrumental role, that is, it is important for them to be healthy because it is a prerequisite for the realization of other goals, for example, to lead a life full of intense athletic activities. But health does not need to be important for all. Since extending the healthspan is important for most people, this objective should also be given greater political consideration than is currently the case. This insight also illustrates why the existence of public universal health coverage is a fantastic achievement. In *We Have Always Been Cyborgs*, I described many current challenges associated with public universal healthcare, such as the question of financing health insurance and, related to this, our handling of personalized digital data.

Here, some fundamental differences between classic transhumanism (CT) and Euro-Transhumanism (ET) have already been highlighted. However, you find a slightly more detailed sketch of the various approaches to transhumanism in the following paragraphs. Transhumanism was coined by Julian Huxley in 1951, whose thinking stands for foundational transhumanism. FM-2030,

Max More, and Natasha Vita-More were responsible for turning it into a broader cultural movement during the 1980s and 1990s. Their approach represents original transhumanism. Nick Bostrom has been responsible for developing a philosophy of classic transhumanism since the beginning of this millennium which is part of the Anglo-American cultural tradition. The most famous proponents of classic transhumanism alive include Elon Musk and Peter Thiel. The roots of Euro-Transhumanism, on the other hand, can be traced to the 2009 publication of my article 'Nietzsche, the overhuman, and transhumanism', which has resonated with artists (Joanna Grochowska, Sven Helbig), entrepreneurs (Martine Rothblatt), and intellectuals (see Tuncel 2017), who affirm transhumanism but whose thinking primarily bears constructive interferences with the continental European cultural tradition from Heraclitus via Spinoza and Nietzsche into our times. To stress the close affinities between Euro-Transhumanism and many of Nietzsche's reflections, I have added references both to Nietzsche's writings as well as to secondary literature on Nietzsche's reflections, so that the complex entanglement between Euro-Transhumanism and Nietzsche's thinking can be grasped and thus stimulate further reflections on many central transhumanist issues.

Some elements of this way of thinking can already be found in thinkers like James Hughes and Martine Rothblatt. However, it is the decisive embeddedness in the continental European cultural tradition which is decisive for Euro-Transhumanism. It must be noted that this relatedness does not imply a eurocentrism or a new type of colonialism, as this approach clearly highlights, stresses, and affirms its own situatedness. It is entangled in a European continental way of thinking while not affirming the universal validity or ultimate superiority of this kind of approach. When entering a discourse, Euro-Transhumanism does not claim a better epistemological rootedness than an alternative account. It merely presents suggestions, highlights insights, and proposes reflections to see whether a constructive or destructive interference occurs with respect to an alternative approach. In addition, it must be noted that a Euro-Transhumanist account could also be affirmed by someone who was born outside of Europe, if that person has been brought up in the continental European cultural tradition or has realized strong resonances with respect to this tradition. Euro-Transhumanism stresses the relevance of glocalized investigations, situated insights, and the entanglement in contingent nodal points.

Core elements of my own take on Euro-Transhumanism were further developed in the monographs *On Transhumanism* (2020), *We Have Always Been Cyborgs* (2022a), and *Philosophy of Posthuman Art* (2022b). The term 'Euro-Transhumanism' was coined in 2022 during an intellectual exchange on the various philosophical challenges which I put forward in the aforementioned three monographs. In 2024, an essay collection entitled *Metahumanism, Euro-Transhumanism, and Sorgner's Philosophy* (edited by

Schussler and Balistreri) was published in which central articles of this intellectual exchange have been collected.

In a nutshell, the following 12 differences between classic transhumanism (CT) and Euro-Transhumanism (ET) can be summarized.

Visions

CT upholds utopias; ET rejects the affirmation of utopias, but suggests visions. In my monograph *We Have Always Been Cyborgs*, I extensively dealt with the issue of utopias. Some classic transhumanists have seriously suggested the death of death, the goal of getting rid of suffering entirely, or a longtermist version of utilitarianism which implies that what matters is merely that some humans survive. None of these issues are either realistic or desirable options. With my German background, I have long been confronted with challenges of utopia. Utopias promise a future which will never and maybe even can never come about, and also have the implication that the quality of lives of currently living people get sacrificed for utopian futures. This means persons are merely treated as a means for a future which will never come about. Such structures must be avoided at all costs. This does not mean that one should not play around with certain visions as inspirational factors. However, it does mean that any utopia which gets taken as a sole universally valid purpose has terrible implications which need to be avoided.

Diversity

CT affirms a hyper-renaissance of perfection; ET affirms a radical diversity of concepts of the good. I dedicated an entire chapter of the monograph *We Have Always Been Cyborgs* to this topic, in which I distinguished between a hyper-humanist account of the good, a commonsense account of the good, and a radically pluralistic account of the good. A hyper-humanist account of the good has been suggested by Bostrom. It basically implies that the Renaissance ideal gets accepted, and technologies get used to realize it, such that the traditional targets can be exceeded. The commonsense account of the good draws upon widely shared elements of the good, like not being ill, not being handicapped, and having highly developed cognitive capacities. It has been suggested by Julian Savulescu. It is tempting to agree with him. However, the challenges come about when the various factors are taken as ultimately justified and valid elements of the good. Then it can have strongly paternalistic implications. I have suggested a radical diversity of concepts of the good as an alternative. It takes seriously that the bundle of drives out of which we all are constituted does not have an essential core element. To increase the likelihood of people living a good life, this diversity needs to be considered. At the same time, I stress that some factors are such that they are indeed widely shared, for example, the relevance of an increased healthspan. It is acknowledged that an increased healthspan is not a necessary element for living a good life, but it is a widely shared element, which means that for most people it is a value which increases their quality of life. The difference between a universally valid concept and a widely shared one is decisive here, which is also a decisive element of twisting goodness. A concept of the good gets presented, but it is not comprehensive, universally valid, as well as unconditional. Thus, it is a twisted concept of goodness.

Pessimism

CT has an optimistic worldview; ET regards Dukkha as plausible, and consequently a version of philosophical pessimism. Some classic transhumanists claim that a world without suffering can indeed be realized, and that minds can be uploaded soon, such that we can experience lives full of the most wonderful types of fulfilment. This is wishful thinking. I am not even certain whether such visions are desirable. However, I do not regard these descriptions of the world we live in as plausible. The world can be interpreted in so many ways. We can tell an enormous number of stories to make sense of the world. Any story can always be retold in a different manner. However, 'some fundamental things apply as time goes by'. One of these fundamental insights is the relevance of suffering, of Dukkha. If the ego wants something, it suffers, and there is no way of escaping this fundamental structure. The challenge lies in how we deal with suffering. This insight gets acknowledged by many world religions. They provide various responses on how to deal with the permanent suffering people have to deal with. Moments of fulfilment are rare, and they are always mixed with pain. Intensely overflowingly fulfilling moments of pleasure–pains are the best we can hope for. These are even more special than simple moments of pleasure. Dukkha is inevitable. It needs to be acknowledged, which I have done in various publications of mine, starting from my book *Metaphysics Without Truth* which came out when I was 26.

Perspectivism

CT affirms the universal validity of certain values and norms, which can have neo-colonial implications; ET regards perspectivism as plausible, which undermines the possibility of neo-colonial implications. Classic transhumanism has suggested so many different norms and values, and they have claimed universal validity for these normative judgements. This is highly problematic for various reasons. One of them is an ontological worry. How and where should any such universally valid normative judgement exist? Mackie's account of values and norms in *Ethics: Inventing Right and Wrong* (1990) is plausible.

Objective values and norms would have to be like Plato's forms, extremely weird ontological entities in a realm which cannot be accessed by means of our senses. I agree entirely. However, if this is not the case, what then are norms and values? They are fictive entities created in the interest of their inventors. By analysing the underlying motives of the inventors, the relevance of the normative judgements can be grasped. If this is a plausible account of normative judgements, the issue of neo-colonialism does not come up either, as neither norms nor values would claim their own universal validity. This does not mean that autonomy needs to be given up, but the implications of a liberal ethics of fictive autonomy need to be thought through.

Hermeneutics

CT affirms a version of utilitarianism; ET embraces a hermeneutic understanding of ethics. Utilitarianism which is so widely shared in the Anglo-American world has enormously rigid implications. The morally right action can be calculated on the basis of a mathematical formula. Appropriate numerical values need to be connected to possible outcomes, probabilities need to be associated, and then the morally right action can be analysed. This way of dealing with ethical challenges might be helpful when dealing with some political issues; however, in many 'real-life' circumstances, the particular circumstances of a situation are so nuanced that a hermeneutical procedure provides more plausible judgements. The devil is in the details. Cultural resonances, intellectual associations, and associative connotations need to be considered. Consequently, it is advisable to always progress with enormous care, when dealing with ethical challenges and moral dilemmas. Hermeneutic reflections by Vattimo demonstrate explicitly how a hermeneutic understanding of ethics can be applied.

Social democratic liberalism

CT argues for libertarianism; ET embraces a social democratic version of liberalism. The political aspect is particularly tricky. I am aware that political norms, too, are human-made fictions. Here, I use the word 'fiction' such that it does not imply that a fiction has to be false but merely as identical to the word 'perspective', which implies that it is a human-made judgement which can be false. The overall political system depends on what representatives of interest groups have fought for successfully. There is not one ideal political arrangement which is the ultimate goal or which is valid for all times. Hence, various advantages and disadvantages need to be considered. Thereby the question of advantages and disadvantages for whom always has to be considered. When it comes to developing new pharmaceuticals, I can see advantages of a libertarian system. However, when it comes to issues like accessibility, then the question needs to be taken seriously, in which kind of society do you want to live? A society in which people do not have access to a publicly funded healthcare system seems to me highly problematic. If you are not feeling well, and you need to think whether you can afford to be taken to a hospital or to a physician, this is a terrible situation. Nation-states are there to take care of the well-being of the people living in these states, and having access to a public health care system is such a fundamental element of most people's well-being that it needs to be covered. It seems even more important than having a public educational system. Given these intuitions, a social liberal political system seems to be an as-good-as it-gets-system when it comes to possible political organizations. There are many challenges associated with it, but it is as-good-as-it-gets for the time being.

Increased healthspan

CT aims for immortality; ET aims for an increased healthspan. There are many challenging concepts which are upheld by classic transhumanism. However, the concept of immortality is clearly among the weirdest. If mind uploading was actually a necessary element of transhumanism, which it is not, then I would understand why some intellectuals accuse transhumanism of merely being a new kind of religion or of it not being a worldview which deserves to get taken seriously by scholars. Immortality, if understood, and taken literally, just does not make any sense in a non-dualistic ontology of becoming whatsoever. If there is permanent change in this world, there is not even any kind of ontological stability which transcends any single moment. If cosmological accounts of the world get taken into consideration, then the absurdity of the concept of immortality becomes even clearer. There is no realistic way in which a person can survive a cosmological singularity. The word 'immortality' works as an advertisement tool; however, it has the side effect that it undermines the intellectual credibility of the proponent. What is relevant indeed is the importance of an increased healthspan. Most people are, in some way or other, interested in living longer healthily. Health here has to be taken as an open concept. For some, health means to feel good, for others health implies the realization of our functional purposes. There are great arguments for both sides. However, when I stress the relevance of health, I do not limit the meaning of the word to a specific philosophical understanding. Health in an extremely general sense is important for most people, and it is this insight which is enormously important. It also has a great number of practical implications.

Carbon-based evolution

CT can be identified with a silicon-based transhumanism which is focused on mind uploading; ET can be associated with a carbon-based transhumanism which has realized that 'We have always been cyborgs', and it is highly unlikely that 'singularity is near'. Many critics of transhumanism have highlighted that transhumanism is all about the longing to become immortal through mind uploading. I have already mentioned a few reasons why immortality is a concept which needs to be avoided. A similar issue applies to mind uploading, even though I cannot exclude the possibility that it will not be possible sometime in the future. However, given current scientific insights, there is no indication whatsoever that silicon-based entities can have consciousness, self-consciousness, emotions, or intentionality. However, in order for it to be possible for our personalities to actually be transferred to a silicon-based foundation, it has to be the case that they can be transferred with a loss. Hence, consciousness, self-consciousness, emotions, and intentionality would have to persist on a silicon-based foundation, too. There is no indication so far that this can actually be realized. This does not mean that it is impossible, but there is simply no good reason to think that it will be possible to realize this goal soon. However, this does not mean that there is a categorially dualistic ontological distinction between silicon-based, and carbon-based entities. Both exist in the sensually accessible realm. It is a matter of structuring which qualities can be realized. With the coming about of the capacity to think, speak and reason, something significant has happened, which took place about 300,000 years ago. A genetic mutation plus human interaction brought about that we have rationality. This is when human beings were realized. This upgrade has generated the coming into existence of human beings. Now, it seems plausible that we ourselves can generate further upgrades. However, upgrading ourselves is nothing morally problematic. It rather is the basis of our existence. What needs to be considered is which directions are desirable. Still, there is no general response to this question. Again and again, we need to talk about, analyse, and reflect upon these issues in order to guarantee that the developments are in our interest.

Digital panopticon

CT is centrally concerned with the risks of superintelligence; ET is doubtful concerning the coming about of artificial general intelligence as well as superintelligence in the near future. I do not underestimate the relevance of AI. Playing around with various AI tools is amazingly astonishing, and they are getting better day by day. I can already see some of the developments which are likely to be available in the near future. However, this does not mean that embodied AIs will have to take over the world soon. The human–AI interaction has an enormous amount of potential for being risky for our existence. We do need to be enormously careful in this respect. This does not mean that there is a danger of AI taking over the world. There is a fine but decisive difference between these two judgements. If consciousness, self-consciousness, emotions, and intentionality cannot be realized on a silicon basis, then it seems unlikely that embodied AIs will be able to say 'I want this or that' and realize it. However, such a decision would have to be necessary for AI to actually be in control. Such thoughts are great for utopian novels, sci-fi films, and Netflix series, but AI taking over by itself is not a realistic option so far. Challenging moral issues arise concerning human–AI interactions, and the coming about of the digital panopticon. However, this phenomenon is fundamentally different from AI's taking over by themselves.

Creating a meaningful lifestyle

CT is concerned with the overcoming of suffering; ET recognizes the inevitability of suffering. In contrast to the former statement concerning Dukkha which was a descriptive one, here, normative judgements get dealt with. Classic transhumanists claim that we should try to overcome suffering, as suffering undermines the capacity for living a good life. Should we actually do so? Suffering is a challenge. We do not like to suffer. However, it also needs to be realized that we usually embrace suffering when it is meaningful suffering, when we know what we suffer for, when we suffer to realize something which is important for us. Creating a sculpture, writing a book, or climbing onto the top of a mountain can be extremely meaningful activities for many humans. These activities can provide lives with a purpose. They involve suffering, pain, and discomfort. However, these side-effects get embraced, because of the goals which are being aspired for. Hence, suffering does not have to be a problematic issue. It is meaningless suffering which is so terrible. A second point needs to be considered, too. If we imagine a life without suffering, the issue also needs to be addressed in how far pleasure can be experienced in such a world. If an intense feeling of pain is not possible, how and why should the feeling of pleasure be possible still? This does not seem to be the case. The question then needs to be addressed whether a life without pleasure and pain would indeed be better than one with pleasures and pains. Buddhists might think that this is the case. I have my doubts in this respect.

Ambiguity of the Enlightenment tradition

CT sees itself in the tradition of the Enlightenment; ET is highly critical of the logocentric Enlightenment tradition and stresses an ambiguous relationship with the philosophical foundations of the Enlightenment. This point is particularly intriguing. What does the Enlightenment stand for? There are two strong traditions: (1) the logocentric tradition from Descartes to Kant and (2) the body-centric tradition from de Sade to Nietzsche.

Depending on your understanding of what enlightenment is, different ontological implications are given. According to the logocentric tradition, human beings consist of a physical body and an immaterial rational mind. Enlightenment here is about using reason to criticize traditional institutions, structures, and thoughts. According to the body-centric tradition, however, enlightenment is about undermining the universal validity of reason to realize that humans are merely gradually different from other animals, and that our rational mind is merely a part of our body. In this case, however, there is no longer one unified rational mind, but there are as many minds as there are bodies, and each rational mind is slightly different from the mind of another body. Classic transhumanists embrace the understanding proposed by the logocentric tradition of enlightenment that it is our task to use reason to realize our goals, but at the same time many of these transhumanists also affirm that they embrace a naturalist ontology. Such an ontology, however, rules out the possibility of categorically dualist ontology as it is affirmed in the logocentric tradition from Descartes to Kant. Hence, it can be doubted whether such a philosophical approach can consistently be upheld. Euro-Transhumanists take this insight seriously and embrace the understanding of an embodied mind. This leads them to highlight an ambiguous relationship with enlightenment. On the one hand, the use of reason is important, helpful, and relevant. On the other hand, it becomes clear that reason can no longer provide us with fundamental insights concerning the nature of the world. If the mind is merely a part of the body, then it is plausible that all it can do is to somehow help us navigate through this world without there being the option of reason providing us with an insight into the nature of the world

Nietzsche is an ancestor

CT denies that Nietzsche is an ancestor of transhumanism; ET regards Nietzsche as an ancestor of transhumanism. Last but not least, I will mention a few additional points concerning the relevance of Nietzsche to transhumanism. A central reason for classic transhumanism to reject Nietzsche as an ancestor of transhumanism is that Nietzsche still has an enormously bad reputation, as he is widely seen as being intimately connected to the so-called Third Reich. In order to protect transhumanism from a bad reputation, classic transhumanism distances itself from getting associated with Nietzsche. There are other reasons for doing so, too, but the fear of classic transhumanists of getting a bad reputation for being associated with Nietzsche seems to be among the most dominant reasons for upholding this judgement. For Euro-Transhumanism as well as for informed scholars of Nietzsche's philosophy, it is entirely clear that Nietzsche is a good European, an anti-German, and an anti-anti-Semite. The collection *Nietzsche and Transhumanism*, which was edited by Yunus Tuncel (2017), provides an excellent overview, and many leading Nietzsche scholars have contributed informed and insightful articles to this collection. In contrast to classic transhumanists, Euro-Transhumanism draws heavily on reflections from Nietzsche, is deeply engaged with his reflections, and employs Nietzsche's thinking for developing further Euro-Transhumanism. There are many structural analogies between Nietzsche's philosophy and Euro-Transhumanism. The following three insights seem to me of particular relevance: (1) Nietzsche highlights that the soul is merely a part of the body (KSA, Za, 4, 39); (2) Nietzsche's will to power ontology of becoming is a non-dualistic version of naturalism (KSA, NF, 12, 312f; Sorgner 2007); Nietzsche's ontology of becoming can be seen as a reinterpretation of the ontology of becoming as it was presented by Heraclitus (KSA, NF, 12, 211, 312f, Sorgner 2022b); (3) Nietzsche also highlights the relevance of the wisdom of the body with all its idiosyncratic drives, wishes, and longings (KSA, Za, 4, 39). There is a lot more which could be said on this issue.

Transhumanism is thus about moving beyond our current limitations, whereby each person should have the right to realize their own respective objectives.[6] Here I always emphasize the radical plurality of the good, which is why the legal norm of negative freedom is an outstanding achievement, since with this kind of freedom comes the absence of coercion. Transhumanists usually highlight the importance of morphological freedom. Reproductive freedom and educational freedom are also manifestations of the norm of freedom (Sorgner and von Becker 2023). However, these statements do not imply that transhumanism must be libertarian in orientation, that is, that only negative freedom plays a role. This understanding only applies to classic transhumanism. Most serious academic transhumanist approaches, including Euro-Transhumanism, affirm social democratic variants of liberalism. An increase of the healthspan usually plays a central role. For this to be realized, according to many individuals, there must be restrictions on freedom, if only in the form of compulsory payment for public health insurance. If negative freedom is emphasized exclusively, then it undermines itself, as the tensions between the socially distinguished and those to whom this does not apply otherwise become too great. Hence, freedom and democracy are incompatible, if freedom solely stands for negative freedom. Thiel is correct in this respect.[7] However, he stands with negative freedom, whereas I affirm that a plausible understanding of freedom has to include positive as well as negative aspects. Thus, freedom includes a certain basic financial security, especially with regard to the question of health. Numerous other demands go hand in hand with this insight. Some facets of the diversity of implications of this basic attitude will unfold during the considerations presented here. In any case, this aspect is a central philosophical challenge for classic transhumanism.

Transhumanism aims to promote the quality of life for all persons, which can usually be done by means of technologies. The concept of technology should be understood in a broad sense. It includes practices of body shaping as well as the use of external aids. This means that learning a language, just like using a telephone, is a technological process. With the help of such technologies, personal boundaries can be transcended significantly such that posthumans can emerge. It should be emphasized here that transhumanists affirm a great variety of conceptions of the posthuman. Basically, a distinction can be made between silicon-based transhumanism, usually affirmed by classic transhumanism, and carbon-based transhumanism, which is being upheld by Euro-Transhumanism (Sorgner 2020). It also needs to be noted that classic transhumanism affirms a strong and detailed Renaissance ideal of perfection, while euro-transhumanism highlights the challenges concerning any non-formal concept of the good.

Representatives of silicon-based transhumanism, characteristic of classic transhumanism, assume that the posthuman must be a digital entity. Its emergence can take place in different ways. An algorithm may independently evolve to produce the posthuman. Another option is that the digitization of a personality through brain–computer interfaces, such as Musk's 'Neuralink', will lead to the emergence of posthumans. Ray Kurzweil is certainly the best-known transhumanist in this regard. Euro-Transhumanism regards a carbon-based further development as more plausible. Representatives of carbon-based transhumanism, on the other hand, tend to identify the posthuman with an organic being that will emerge by means of gene technologies or brain-computer interfaces, whereby it remains an open question whether the posthuman will still belong to the human species or not, that is, whether a posthuman will be able to reproduce sexually with humans or not. Both options are advocated by transhumanists. The only decisive factor is that the currently prevailing boundaries are significantly exceeded.

In general, I emphasize that due to technical innovations, reproduction and sexuality are becoming increasingly disconnected from each other. Condoms, the pill, surrogate motherhood, artificial insemination, and the artificial womb are decisive innovations in this regard. According to a Catholic understanding of natural law, sexuality should always be linked to the possibility of reproduction. Because of various technical inventions, there is the possibility of increasingly decoupling sexuality from reproduction. Sexuality is for personal fulfilment. Reproduction is increasingly professionalized with the help of technical aids in order to increase the probability of a good life. The promotion of quality of life with technical support is a central transhumanist concern.

The resistance to genetic engineering, brain-computer interfaces, and digitalization, especially in German-speaking regions where the legacy of eugenics is identified primarily with the legacy of the 'Third Reich', need not be so strong. This is made particularly clear by the fact that it has only been possible to realize our current high quality of life by means of technologies. Here, I refer not only to technologically advanced regions of the world. Our basic humanity is guaranteed by a technological upgrade. Our ability to speak, which is guaranteed by a genetic mutation, has been further refined through parental and cultural upgrades of upbringing and can be further developed.[8] This process of controlling language ability is a technological process. By means of our linguistic capacities, we are steered organisms, or, in other words, we are cyborgs. A cyborg is, after all, a cybernetic organism, the word cybernetic deriving from the ancient Greek helmsman, *kybernetes*. So, it is not the case that we acquire our ability to use a language by means of an immaterial divine spark that comes from outside of the material world. Rather, the ability to speak is a technological upgrade. This is how a technologically controlled organism that can speak comes into being. This alternative narrative regarding the emergence of the ability to speak has immense cultural, political, and ethical implications. With this alternative anthropology, the traditional dualistic view of humanity, already found in Plato's allegory of the cave and associated with a material body and an immaterial soul and spirit, is twisted (Sorgner 2022b). It is a twisted anthropology. The cultural significance of this transhumanist anthropology is enormous, and it has hardly arrived in the lifeworld yet due to encrusted cultural structures of our categorially dualistic past.[9]

What is crucial at this point is that we have become endowed with language due to technological changes. Technologies are both tools and processes that change our being-in-the-world. If techniques are used appropriately, our quality of life can be improved significantly. These considerations also apply to genetic engineering, brain-computer interfaces, and digitization, which in turn also means that none of these technologies is necessarily morally reprehensible. The only thing that matters is that they are used in our interest. If it is morally appropriate to extend our healthspan by means of sports and diets, then this also applies to the multitude of genetic techniques, if they are intended to promote the same goals. Based on the revised cyborg anthropology, it becomes clear that technologies are nothing alien to our nature, but that we have only become human through technologies. If we also use new technologies in such a way that morally appropriate goals are promoted, then these are to be evaluated in the same way as this applies to traditional technologies such as education. If it is considered neglectful not to use certain traditional technologies on one's own offspring, then this can also mean that the same applies to new technologies which are intended to promote the same goals and which involve comparable risks. This makes it clear that genetic technologies should not be demonized. The risks, objectives, and framework conditions should be evaluated regarding individual technologies. It is this basic attitude that implies openness concerning the possibilities associated with emerging technologies. There are also many reasons to hold on to this openness, especially if we look at the past handling of technologies. As each new technology was developed, warnings were raised about the possibility of the destruction of life. So far, this has not happened. This fact might provide us with a reason for adopting a rather proactive basic attitude concerning the development of new technologies. The relevance of adopting the proactive principle can be seen not least in the significantly increased life expectancy occurring over the past centuries. I am a Euro-Transhumanist, and I can hardly wait for further technological achievements to be realized.

Euro-Transhumanism: *Ex Negativo*

In the *ex positivo* chapter, I pointed out that there are quite a few misconceptions about transhumanism, both among the general public as well as among researchers, for example, regarding mind uploading or immortality. In this chapter, I address 12 characteristics *ex negativo* that are paradigmatic for Euro-Transhumanism, but are not necessarily to be identified with all types of transhumanism.[1] In this way, some further misunderstandings should be cleared up and the plausibility of transhumanist thinking gets strengthened further. I strive to always distinguish between the prevailing basic attitude (classic transhumanism) and my own variant of Euro-Transhumanism, if this is necessary.

Non-anthropocentric

One criticism of transhumanism is that it is anthropocentric.[2] This criticism holds that transhumanists are only concerned with increasing the quality of human life and that it excludes non-human living beings. There may be a few classic transhumanists for whom this accusation applies, such as Steve Fuller (Fuller 2011; 2012; Fuller and Lipinska 2014). This assessment is certainly not predominant, which is particularly evident in the fact that a widespread transhumanist demand wants to guarantee personhood for non-human animals.[3] Most transhumanists are concerned with persons not with humans.

The term 'human being' is descriptive. It stands for members of the biological category 'human being'. Whether they should be respected or not is an open question. The term 'person', on the other hand, is normative. It implies that all those who are considered persons should be respected in the same way. For a long time in Western cultural history, at least all human beings were considered persons. There was a time when God and angels were also considered persons. However, this term 'person' was mostly not applied to other natural living beings. Thus, among natural living beings, only human beings were persons and deserved to be respected in a particularly prominent way.

After Darwin, Nietzsche, and especially Peter Singer, this assessment was revised by many ethicists. Thus, the respect associated with the status of a person should only be attributed to those who deserve to have it. A living being without a nervous system and without a brain cannot suffer, which is why they should not be accorded with any special respect. An adult orangutan, who can recognize herself in the mirror, on the other hand, should be granted personhood. Argentina has adopted this at the legislative level. As a result, an Orangutan was legally required to be released from the zoo.[4] Following this approach, a fertilized human egg would also not be considered a person. This basic attitude is shared by most transhumanists. It follows from this that not all humans are persons, but some non-human living beings should certainly be considered persons. It is therefore not the case that the fact of belonging to the biological human species alone is sufficient for being morally respected. What counts morally is that the corresponding living beings possess morally relevant characteristics, such as a strong ability to suffer. Such a position is not anthropocentric but person-centric. Euro-Transhumanism clearly affirms such a position.

But even this position is sometimes criticized as dualistic, especially by critical posthumanists, since a living being is either a person or not, but in this way, the complexity of life world would not be represented appropriately, since the specific context of a living being would not be taken into account. The variant of person-centred ethics proposed here is not dualistic, since it starts from the capacity to suffer, which entails a further hierarchical ordering of the concept of person. There is a hierarchy of different persons. In addition, a relational approach, as presented by critical posthumanists, is not a pragmatically manageable approach to legal regulation. Furthermore, relational approaches have paternalistic implications, which runs counter to central achievements of the Enlightenment. Crucially, the accusation of anthropocentrism does not apply to transhumanism (Sorgner 2022a).

Non-logocentric

Is transhumanism a logocentric position or not? This question is not so easy to answer. Many transhumanists, like Steve Fuller or James Hughes (Sorgner 2022b), point out that transhumanism belongs to the tradition of the Enlightenment, for which logocentrism is central, since critique by means of reason is the central method of the Enlightenment. Maybe, classic transhumanism bears some traces of logocentrism. On the other hand, many transhumanists advocate for a naturalism that implies the rejection of an immaterial mind. These assessments seem to contradict each other. At the same time, I must point out that there are logocentric transhumanist philosophies, such as Steve Fuller's approach. His notion in this regard is implausible. Since most transhumanists hold to a naturalistic attitude,[5] logocentrism is not prevalent among transhumanists.

Moving away from logocentrism does not imply that we need to uphold self-contradictory assessments. This departure may go hand in hand with a revised assessment of *logos*, reason. Is reason an immaterial entity or is reason something that has evolved, that is, an embodied mind? This is the crucial question. If the mind is an evolutionarily emergent property, then we have no reason to assume any special significance of the mind for cognition of a fundamental truth. If the mind is immaterial and gives us access to eternal truths, then this would be different. In the case of an embodied, evolutionary mind, on the other hand, there is no necessary reason why we should rely on reason for knowledge of truth. Reason then has arisen because it is somehow useful to us and helps us to think or to achieve our goals. Upholding reason, then, is still important. However, the goal is not the realization of truth but the realization of our own goals. This is clearly accompanied by a move away from the formerly fundamental importance of the mind. It is therefore no longer a logocentric position. This judgement is affirmed by Euro-Transhumanism.

Furthermore, it should be noted that this view also undermines the concept of a single unified reason. If reason arises evolutionarily in each of us in a completely idiosyncratic way, then there cannot be one unified reason but as many different reasoners as there are living beings with reason. At the same time, this does not mean that there are no similarities between our rationalities. If I order a cappuccino with cheesecake in a café, then I will also receive this in regular circumstances. This means that, quite pragmatically, a successful and reasonable communication is taking place. Whether what I imagine as a cappuccino can be identified with what the waiter imagines is a completely different question. The only decisive factor is that communication works pragmatically. Using one's own reason is helpful in planning one's life. Whether it is of importance to us in terms of recognizing a fundamental truth about the world, on the other hand, is a completely different question.

Non-essentialist

As a rule, transhumanism does not represent an essentialist worldview. Many transhumanists regard a version of naturalism as plausible. The processes in this world can be characterized by Heraclitus' constant flux rather than with some unchanging essence.[6] The world can plausibly be described as a constant becoming in all respects at every moment. Everything that appears to us as a constant being is only a contingent nodal point that cannot be characterized as unchanging on closer analysis.

There are also other concepts that can be identified with naturalism. The soft naturalism advocated for by Jürgen Habermas implies that most aspects of the human being can be analysed empirically, but he assumes that a small area of the human being lies fundamentally outside the possibility of empirical analysis (Habermas 2004, 877). In this way, the belief in the immaterial mind, which is particularly widespread among philosophers in the German-speaking world, returns under a new guise. To me, this assessment seems highly implausible. Why should I assume something that is in principle outside the possibility of empirical access? How could I ever have access to it myself? Such an anthropology is essentialist. The essence, the being of the human being, then lies in its unchanging, immaterial nature, its reason.

However, there is no strong reason to assume such an entity. I have never encountered this immaterial world. Why should it be postulated? Further, there are numerous violent structures associated with such an essentialist world, such as good and evil. For this reason alone, essentialisms are to be characterized as dangerous as well as implausible.[7]

If we move away from such an immaterial world, what remains is constant becoming. With regard to the human being, we are then left with a living being with unclear limitations (Sorgner 2022b): (1) first of all, the human spirit has become an evolutionarily emergent and thus multifaceted one, that is, an embodied mind; (2) furthermore, it has been realized that a human being consists of 39 trillion non-human cells and only 30 trillion human ones. It thus comprises more non-human than human cells; (3) it is an open question whether the smartphone counts as an extended mind to the human body or not. A person with a cochlear implant, for example, can control his or her own hearing with a smartphone. Is it a violation of her bodily integrity if it is confiscated by a policewoman? (4) Is it a violation of a person's bodily integrity if he or she controls an avatar in the metaverse, which is then touched indecently against the person's will by another avatar?

Perhaps even more groundbreaking are the consequences of non-essentialist thinking for ethics, since there is, arguably, no longer any convincing possibility of ethical truths, that is, good and evil. Norms and values are fictions made by persons who conceive them in their own interest. There is no epistemological superiority of any values and norms. With regard to the claim of ultimate justification, it cannot be established why one set of values should be superior to another. This is affirmed by Euro-Transhumanism, but not by classic transhumanism.

However, this is not to say that values and norms are irrelevant. They can still be contested. Which one should be effective? Which values and norms should prevail? There is no better in terms of any eternal foundational ontological standards. But there is still a better and a worse in terms of our interests. We should not harm an innocent person. Fortunately, there is consensus about this today. If we agree on values and norms, then these are not better founded on an epistemological basis. But they can become effective if we work to promote, establish, and enforce them. Our values and norms may be fictions, but if we want them, they can become effective, active, and powerful.[8]

This understanding likewise prevents the emergence of violence caused by essentialist universalisms, for example, against universalist judgements such as: 'Homosexuals are sick because our natural sexual proclivities are always aimed at procreation.' Such an essentialist judgement arising from a version of natural law involves an enormous amount of violence against individuals. The same is usually true of general essentialist value judgements. To affirm the world of constant becoming also means rejecting violence based on essentialist values and norms.

Non-perfectionist

Transhumanism often speaks of enhancement. Mostly, transhumanism is portrayed in the media with Zeus on Adderall and Hera on Retinol being the valid transhumanist ideals. It must be admitted that the classic transhumanist Nick Bostrom finds it regrettable that we cannot all become Renaissance geniuses. According to him, the Vitruvian Man can be realized through the latest technologies. There is a tendency towards perfection among classic transhumanists (Sorgner 2022a).[9]

For Euro-Transhumanism, these ideas of perfection are implausible. Earlier I mentioned the concept of the radical plurality of the good, from which it follows that nothing can be said about the content of a successful life.[10] The best possible answer in terms of a capacity that increases the quality of life for many people is a prolonged healthspan. But even this cannot be universally valid applied for increasing the quality of life.

If we want to know what people consider a successful life, we should ask them. This does not mean that everyone necessarily knows what lifestyle is in their own best interest, as our cultural environment can sometimes influence us to such an extent that more fundamental needs are obscured. Nevertheless, this can give us an overview of important facets of people's diverse concerns. This is not to say that every desire is morally appropriate. If a person directly causes harm to another person against his or her will, then it is certainly an immoral act, as in the case of rape. However, there is a vast rainbow-coloured kaleidoscope of interests, desires, and appetites. A hike along the lakes and up the peaks around Sils Maria may be a fulfilling activity for one person. Others need the fast pace of a big city like London to be filled with *joie de vivre*. Others have a strong desire for complex historical culture, like Florence, to feel satisfied. Others require the technological pulse of Taipei to justify their existence. It is a radical plurality of needs that goes hand in hand with a high quality of life. Which is why any simple conception of perfection is surely too reductionist to be a plausible philosophical position. If we want to speak of perfection, then at most each person may embody a perfection of their own. However, this is a far cry from a traditional perfectionist ethic that assumes that a single ideal of a perfect life is to be identified with the successful life.

This understanding of human needs is also the reason why morphological freedom plays such a central role in transhumanism. With the help of technological processes, numerous concerns, goals, and needs can be achieved much more easily, efficiently, and quickly than would be the case without them. By means of a website, even *Elisenlebkuchen* can be ordered much more conveniently than was possible just a few decades ago. A trip to a remote mountain village in the Alps can be easily planned from home these days. It may be difficult to find like-minded people in your own village. Using social networks, such contacts can easily be realized all over the world.

Non-paternalistic

To what extent can global challenges like global warming be addressed with a focus on freedom? Is transhumanism about the attitude: *après moi, le déluge*? The focus on individual freedoms is about avoiding paternalism. Totalitarian structures should be avoided at all costs, as they undermine the diversity of human flourishing. If this is the case, how can sustainability be taken into account?

What are our alternatives? Let us assume that global warming is the most important current challenge. Then, in one country, appropriate laws could be passed. Ban meat eating, flights and – last but not least – reproduction, as population growth is the main cause of the increased emission of carbon dioxide. Every additional human being born represents a new problem for climate change or, more precisely, global warming. Other processes associated with carbon dioxide emissions are also closely causally linked to population growth. More people, more energy, more food. Anyone who sees global warming as the central current problem should feel morally obliged not to bring about a new generation. If a nation-state starts from this basic assessment of global warming, then it would have to limit reproduction in the same way that, for other reasons, China implemented its one-child policy – at least for a certain period of time. Of course, if one country passes such laws, this does not mean that other countries will follow, let alone that analogous regulations will be adopted worldwide. An effective approach must look different.

An alternative approach establishes sanctions to tax processes responsible for global warming, such as higher taxes on meat, flights, and *fincas*. This approach represents a challenge to social justice. Meat, flights and *fincas* only for the rich.

Both approaches are morally problematic. Prohibiting procreation is a totalitarian act. Taxing procreation undermines social justice. Neither approach is an efficient solution because it cannot be easily applied to the whole world. Fortunately, we do not have a world government. The inhabitants of China and India already make up a third of the world's population. With increasing wealth, which can hardly be stopped in China in particular, comes the desire for higher meat consumption. We do not even need to talk about increasing energy consumption. Developing countries depend on energy to promote their own prosperity. 'You in the West have created a high level of prosperity through the consumption of energies over the past 200 years, now we have the right to do the same.' Advertising and moral appeals will certainly not bring about a radical change in these countries. Thus, it is necessary to rely on other approaches.

Bans are totalitarian. Taxes may be socially unjust. Moral appeals are ineffective. Max Roser, an economist and statistician at Oxford University, who developed the fantastic website ourworldindata.org,[11] has shown that population development proceeds in four stages. Qualified medical care and improved hygiene reduce infant mortality and increase life span. A further reduction in the reproductive rate can be promoted through improvements in education, an improved quality of life and secure housing, food, and social security. The reproduction rate in Germany is 1.5, in Austria 1.4, and in Italy only 1.2. Analogous processes can also be observed globally. If the quality of life, hygiene and access to education, healthcare, and basic food continue to rise as they have so far, studies by the United Nations suggest that the 12 billionth person will never be born. If this research is correct, it would mean that overpopulation would never be a problem, because innovations can improve the quality of life in such a way that the reproductive rate drops significantly.[12]

These considerations lead us to the departure point for transhumanist reflections on freedom. Relying on freedom and technical innovations does not necessarily imply the basic attitude: *après moi, le déluge*. Only in this way can a sustainable approach to global challenges such as global warming be taken efficiently and with due regard for important cultural achievements. Through innovation, education, and hygiene, the quality of life can be improved worldwide. At the same time, the decline in the reproductive rate can be brought about without the need to adopt totalitarian and socially unjust measures. A non-paternalistic transhumanist approach is not a problem for dealing with a global challenge such as global warming but rather a promising solution. Such solutions are affirmed both by classic transhumanism as well as by Euro-Transhumanism. Some differences might apply when it comes to the financial regulations which ought to be taken into consideration.

Non-totalitarian

Another accusation against transhumanism is that it is libertarian and thus totalitarian, if not fascist. This consideration is based on the following train of thought. Totalitarian structures are not only found in authoritarian systems, such as China, but also in libertarian systems, such as the US. Due to the weak or almost non-existent social security in that country, there is an enormous tension and a large gap between the rich and the poor. The poor are forced to accept work to survive, which is incompatible with a dignified life. Transhumanism, in this interpretation, endorses exactly such a libertarian order and is thus responsible for totalitarian social structures. It is not the case that this accusation is entirely unfounded. Classic transhumanists like Peter Thiel, Zoltan Istvan, and Elon Musk affirm libertarian thinking, with which such challenges certainly go hand in hand. But this need not imply that such a political ideology applies to all transhumanists. Most scholarly transhumanists affirm social-democratic variants of liberalism, which I do not distinguish from, and also includes classical liberalism (Hughes 2014). This also applies to Euro-Transhumanism.

As far as democratic regulations are concerned, however, we must rethink currently dominant structures in relation to technical innovations. This will be briefly illustrated by the example of the importance of digital data. At present, we are striving to protect privacy regarding digital data in a democratic order. In principle, this is a laudable endeavour. However, given the importance of digital data, it is not a practically realistic option if we want to maintain our quality of life in Europe. Central to this is the realization that information flows now as oil once did – the lifeblood of empires. The relevance of digital data can hardly be underestimated against the background of current technical developments. There is no area of life for which research with digital data is irrelevant. Research with personalized digital data is indispensable for policy making, technical innovations, engineering sciences, social and behavioural science as well as natural science research, and especially also for the large field of medical technologies.

If we focus only on privacy, as is currently happening in Europe due to the General Data Protection Regulation, then the possibility of comprehensive and efficient collection of and research with personalized digital data will be undermined. If data are indeed the new oil, then this approach will have a significant impact on Europe's economic prosperity, as digital data are handled differently in other regions of the world. In China, companies are required to share digital data with the government. In the US, companies are allowed to collect, exchange, and sell data without major obstacles. In both regions, it is possible to collect big data and conduct research with it. In China, the collection process is even more efficient due to more authoritarian structures. This has financial consequences, but also implications for research, innovation, and policy making. Money can only be spent once, so it stands to reason that it will flow even more to the US or China in the future than it already does at present. As a result, those of us in Europe will not be able to maintain our current standard of living, which in turn will have an impact on our social security, from unemployment insurance to health insurance. First, it will probably affect the middle class, which will realize that it is, if not absolutely, at least relatively worse off than the corresponding class in China. Whenever this is the case, blame is sought, mostly found in foreigners, refugees, or other vulnerable populations. This leads to social tensions that can go as far as civil war. This is what the future of Europe will likely look like if digital data are as relevant as widely assumed, while the European Union's (EU) Basic Data Protection Regulation will remain in place. The only hope for us may be that Europe, as the new Switzerland, will become the refuge of the world's billionaires, so that money will come to Europe in this way. But this option is probably not realistic (Sorgner 2022a).

In order to continue to maintain our standard of living, it is important to rethink our approach to the use of digital data. We need personalized digital data. These need to be collected to remain globally competitive. In China, the authoritarian state intervenes to uphold the prevailing ruling structures. This process cannot be reconciled with Enlightenment achievements. In the US, on the other hand, it is primarily big tech companies that collect digital data to maximize their profit. This attitude is also affirmed by classic transhumanism. We think tech companies are providing us with goodies, but we are their products or their unpaid workers. In both structures, there is no democratic use of our digital data. Instead, data should be used in our interest. For this to be possible, an EU-created institution should collect personalized digital data. After all, I have more trust in a democratic government than in a private US company focused on maximizing profits. Of course, here too, care must always be taken to ensure that our data are not misused, which is why it should primarily be processed by algorithms, and humans should only have access to them in exceptional cases. Humans are always corruptible. Digital data can then be passed on, usually depersonalized, so that research can be conducted with it. The purposes for which they may be sold must, of course, always be continually discussed. The money collected should then be used for democratic purposes, so that the compulsory collection of data is not an expropriation but a payment for something that is in our interest in the democratic interest (a tax). The revenue should then be used to cover, at least in part, the costs for a public universal health insurance, since an extension of the healthspan is identified by most people with an improved quality of life. The data would not be used to support domination structures of an authoritarian state or the profit of tech entrepreneurs, but in this case the personalized digital data would be used to finance the universal public health insurance, which is justified because most citizens identify an extended healthspan with an improved quality of life. In this case, there would be an actual democratic use of our digital data. This change requires a rethinking of the meaning of digital data. This is the approach Euro-Transhumanism defends.

In addition, in order to reduce the risk of illegitimate state sanctions, the norm of freedom should be given even greater importance than is currently the case. This means that sanctions should only be applied when a person is directly harmed. Otherwise, the greatest possible freedom should prevail. This proposal has significant implications, for example, regarding the legal position on incest between competent adults. Currently, it is still the case in Germany that incest between competent adults is prohibited. In a liberal state, this should not be the case, as it is exclusively a relationship between competent adults. If they want to be intimate with each other, then the state should have nothing to say about it. At this point, an objection is often raised about the likelihood of incestuous relationships producing congenital diseases. However, this objection is not valid in many respects: (1) incestuous relationships between competent adults could also happen between people where there is no possibility of procreation; (2) if the argument were valid that sex can be prohibited when there is a high probability that persons with disabilities may arise, then it should be mandatory by law to have a genetic test regarding genetic compatibility before a sexual relationship may be initiated. Similar procedures are followed in Saudi Arabia. Before marriage, it is obligatory to have such a genetic test regarding sexual compatibility;[13] (3) If the preceding logic were the decisive one, then sex between people with a hereditary genetic defect should also be prohibited. A sexual relationship between two genetically deaf people would then have to be illegal. Such eugenic regulations are immoral. Moreover, such utilitarian considerations are not in line with our legal traditions. What matters here is the consent between competent adults: we want to be intimate with each other. If all competent adults involved agree to this, then no one is harmed.

The greater the importance of the norm of freedom at the legal level, the lower the risk of an illegitimate legal sanction occurring due to the personalized collection of digital data. That legal sanctions are made is good and necessary. If an innocent person is murdered or someone is raped, then it is right that sanctions apply. The more extensive the digital data collection, the higher the probability that the guilty will be caught and punished. Thus, a rethinking of the use and meaning of digital data has important implications that further increase the likelihood of improving personal quality of life. The benefits of universal public health coverage will be expanded. The possibilities of alternative living options will be increased. The likelihood of crime will be reduced, and the punishment of criminals will increase.

These changes promote non-totalitarian structures in different ways. The financing of public universal health insurance undermines the social divide, which undermines possible paternalistic structures associated with excessive hierarchical property gradients. At the same time, the plurality of the good is further expanded at the legal level, so that illegitimate paternalism mechanisms regarding the conception of the good life are largely undermined. This is also a non-totalitarian revision. This version of transhumanism takes non-totalitarianism seriously and is concerned with undermining and dissolving totalitarian structures. The decisive legal norm is the achievement of freedom, which always acts as a corrective guiding principle (Sorgner and von Becker 2023).

Non-dualistic

A common misconception about transhumanism is that it is dualistic and therefore only a humanism on steroids or a hyper-humanism (Ranisch and Sorgner 2014; Tuncel 2014[14]). Crucial to this judgement is the view that mind uploading is central to transhumanism, which is inaccurate, and that this process would reveal a dualistic ontology already present in monotheistic religions, which is clearly false. Just as the immortal soul is dualistically separated from the material body, transhumanists would assume that the personality can be detached from the body to upload it onto a hard drive. This analogy is inaccurate. What is decisive is that the traditional religious soul–body duality is an ontological one. The body is material and the soul immaterial, that is, inaccessible to empirical observations. This is not the case concerning most transhumanist approaches. The personality and the body are located within the empirically accessible world, even in the case of an uploaded personality on a hard drive; if this were possible, it would not be an immaterial entity. It is therefore precisely not a two-world-ontology, that is, one that assumes a material and an immaterial world, that is, a categorical ontological duality. This is a decisive difference.

Whether mind uploading is possible or not is a separate question. In contrast to classic transhumanism, Euro-Transhumanism is rather critical of mind uploading, at least as far as its realization in the foreseeable future is concerned (Sorgner 2022a). However, many of the questions associated with mind uploading are philosophically intriguing. Here is just a small selection of relevant considerations.

One reason why people want to have themselves cryonically preserved, that is, frozen, after their death is the following. In principle, there is a possibility that, due to advances in medical technology, the reason for death can be reversed in the future. Let us say a person was diagnosed with total brain death and his head were cryopreserved after his death. After a few decades, we have succeeded in reversing brain death due to bioprinting, synthetic biology, and nanobots. The destroyed organic structures could thus be restored. Then total brain death would not be accompanied by an irreversible loss of consciousness. If it were also possible to download the personality to a hard disk, for example by means of brain–computer interfaces, then a person who has already died of brain death could be revived. Without cryonics, life on earth is irreversibly lost. With cryonics, there is hope, albeit very slim, of continued life in the future.

I admit that all these ifs are rather questionable given currently prevailing medicine, especially since, in the latter case, such a survival would take place on a silicon basis, and we do not know of any silicon-based living being at present. Stephen Hawking would object here that the computer virus is already a life form.[15] Since the computer virus lacks a metabolism, I am rather sceptical about this assessment. Nevertheless, it is worth noting that it seems to be possible in principle for life to arise from something inorganic. This is exactly what happened on Earth about 3.5 billion years ago when the first life forms, such as bacteria, emerged on Earth. Given suitable boundary conditions, life can also arise from something inorganic. It is therefore not impossible that life is also possible on a silicon basis.

It was important for me to briefly describe these considerations here. Decisive current challenges with regard to transhumanism are different, since it seems highly unlikely to me that the processes just mentioned can be realized in the next few decades. At this point, it is only crucial for me to emphasize that most transhumanists do not affirm a dualistic anthropology.[16] Euro-Transhumanism regards a non-dualist ontology of permanent becoming in all respects at all times as most plausible (Sorgner 2022b).

Non-aletheic

Aletheism is a type of morally illegitimate discrimination that is still widely neglected in critical discussions. It is as reprehensible as racism, sexism, and speciesism. However, it concerns an even more fundamental intellectual level. Here, there is morally illegitimate discrimination against all those who disagree with the one truth that corresponds with reality. So any dogmatic worldview with a strong claim to truth is aletheic, since it assesses all those who disagree with it as ignorant. This assessment has immense implications, especially in the field of ethics (Sorgner 2022b).

In contrast to classic transhumanism, Euro-Transhumanism is non-aletheic. It assumes an ontology of constant becoming in every aspect in all moments. This understanding alone makes it clear that any correspondence of judgements to reality is fundamentally unconceptualizable, since rigid words cannot correspond to constant becoming. From this follows the affirmation of a form of perspectivism, that is, epistemology, that every judgement regarding fundamental worldview statements is only an interpretation.[17] However, having an interpretation does not mean that it is a false judgement, but it solely implies that the judgement can be false. Of course, this also applies to perspectivism itself. In order to show that perspectivism is a false judgement, it would be necessary to show a true insight regarding fundamental worldview statements (Sorgner 2007). Since, in my view, this has not been achieved in the nearly 3,000 years of philosophical history, perspectivism is the most plausible epistemology compatible with an ontology of constant becoming. The ontology of constant becoming suggests perspectivism, and perspectivism makes an ontology of constant becoming seem plausible (Sorgner 2007). This is what I refer to as twisting truth, as thus an ontology can be presented, but without claiming that it corresponds to the world. Perspectivism can be refuted as soon as someone succeeds in demonstrating true knowledge about fundamental worldview statements, which, however, has not yet been realized.

It should be noted here that mathematical statements, such as that 2 and 1 equal 3, are also not truths. Here only a tautology is to be found. Nothing is said other than 3 equals 3. Whether the numbers 1, 2, or 3 have anything to do with reality is a separate question. It seems much more plausible to assume that numbers are abstractions that are helpful in the living world, but which are not encountered in an identical manner in reality.

Furthermore, it should be emphasized that a perspectival epistemology does not imply that scientific insights are irrelevant. They are just not fundamental worldview insights. However, this does not mean that they are not useful. The opposite is the case. Scientific insights are pragmatic truths that help us to survive in the life world. Even the greatest critics of the natural sciences often commit a performative self-contradiction, for example by getting into an aircraft. By flying, they performatively express their trust in scientific insights. They would not do so if they assumed that the plane would not land safely at its destination. But since this is reliably the case, it becomes clear that scientific insights are pragmatically helpful. However, they are not ultimate justifications regarding the nature of the world, with which aletheism would go hand in hand, which makes it clear that transhumanism is not aletheic, but it has good reasons for relying on scientific insights.

Non-utopian

Transhumanism is often accused of being a utopian worldview. There are undoubtedly numerous transhumanists who affirm utopias, for example, Nick Bostrom (2008), David Pearce (2021), and Ray Kurzweil (2014). Euro-Transhumanism regards this basic attitude of classic transhumanism to be highly dangerous. It is important for me to point out that a plausible transhumanism must be non-utopian. Utopias are dangerous, be it the utopia of a classless future society, the utopia of bodily resurrection at the end of days, or even the utopia that we will live as uploaded personalities in a digital society controlled by a benevolent superintelligence.

The central danger of all utopias is to sacrifice the present in order to realize a perfect future social order that can never be implemented in practice.

Here the lives of present people are sacrificed to a future cloud cuckoo land, which is dangerous, and which undermines what is the central transhumanist concern, the promotion of the quality of life of all persons.

When Euro-Transhumanism affirms the democratic use of personalized digital data, no utopia is thereby affirmed but only a vision that has been formulated based on current insights. It is an as-good-as-it-gets vision that has no necessary validity but represents a contingent draft, proposal, and reflection on the basis of which further discussions can take place. None of the reflections presented here represents a necessarily valid final insight. Every reflection can and should be revised, twisted, and reformulated given the possibility for better insights. It is this basic attitude that distinguishes contingent visions from grand, perfect, and ideal utopias.

Utopias have to do with ultimate justifications. Visions are as-good-as-it-gets proposals that invite further reflection and discourse. It is this openness to alternatives and better scientific insights that is characteristic of Euro-Transhumanism. Utopias are opposed to such an objective and basic attitude. All worldviews that affirm utopian goals are dangerous and should be rejected and resisted. A non-utopian basic attitude is accompanied by the possibility of contingent visions, which, as proposals, can represent foundations for further considerations. This is also how the considerations presented here should be understood.

Non-foundational

Habermas' most recent accusation regarding transhumanism was that it resembles the 'worldview of a cult' (Habermas 2014a). The preceding considerations show that this accusation is inaccurate. With a naturalism in the transhumanist sense, in which a constant becoming is assumed, precisely the possibility of an ultimate justification cannot even be thought consistently.[18] A cult, however, usually pretends to do just that. A cult has the answers to all our ultimate questions and challenges. Such a concern is precisely not represented by transhumanism. Rather, it is about pragmatically increasing the likelihood of advancing the quality of life for all people. Our experience suggests that by transcending our current limitations through technology, this goal is likely to be realized. Such a cautious, scientific, and critical approach is not a characteristic of a cult.

Cults demand submission, the following of rites, the practice of certain prayers and the valuing of certain symbols. There is no organization in transhumanism from which one can be excommunicated, nor are there prayers, initiation rites, and sacred symbols that transhumanists must follow. Often, cult members are required to pay membership fees. There are no such things to be a good, practising transhumanist either.

At most, being a transhumanist is associated with the view that technology can be used to increase the likelihood of people living good lives. Politically, religiously, and culturally, transhumanists have a wide variety of different backgrounds. That plurality plays a crucial role in transhumanism is also evident in the central importance of morphological freedom. Every person should have the right to be happy according to her own *façon*, as long as she does not limit the freedoms of other persons in the process. What this means in life-world terms must always be analysed, thought through, and discussed anew. Since the assessments that prevailed yesterday may no longer exist tomorrow. Transhumanism represents an approach to develop contingent guidelines for a constantly changing world and to make the constant becoming worth living. This is by no means an easy challenge. Constant becoming is accompanied by continuous suffering. This Buddhist pessimism is widely upheld among transhumanists.[19] At the same time, however, transhumanism advocates positivity, assuming that our quality of life can be significantly improved with the help of technical, cultural, and social innovations. How the necessary interventions should be designed must always be discussed anew. Changing boundary conditions and insights requires revised measures. This makes it clear that transhumanism is not a sect with an ultimately justified worldview. This accusation by Habermas is probably an attempt to philosophically discredit transhumanism. I hope that the reflections presented here make it clear that transhumanism is to be taken seriously philosophically, and that not all transhumanists have to be philosophical lightweights.

Non-heteronormative

In natural law, it was often argued that every part of the body had its own goal, and that the natural goal of the genitals is reproduction. Consequently, genitalia would only be used in a natural way when the possibility of reproduction is given. On this basis, the demand for heteronormativity emerged. Even though transhumanism is often portrayed in public as a man becoming Superman and women becoming Wonder Woman with the use of technological aids, this does not mean that this characterization is an accurate one. In this case, the typical male and female characteristics would remain. However, their realization could now also be advanced with technological aids. Sometimes Bostrom's version of classic transhumanism seems to have these implications.

This characterization of transhumanism is certainly inaccurate if it is meant to say that Superman and Wonder Woman represent universal ideals and that we should all be guided by them. However, if there is an individual desire to be transformed into a Ken or a Barbie, this need not necessarily be morally reprehensible. What is decisive is morphological freedom, and thus also the right to shape one's own body in the way that is in one's own interest. Can one's own decisions be morally problematic? If a nine-year-old girl is oriented towards manga-inspired ideals on Instagram, feels uncomfortable in her own body, and wants to have appropriate corrections made with the help of cosmetic surgery, then this desire should possibly be questioned, as she is probably not yet competent enough for realizing which needs correspond to her core drives. With competent adults, the situation is different. Their own desires, longings, and dreams should be taken seriously, and our desires are manifold. Just because a notion does not correspond to the socially prevailing one does not mean that the expressed desire need to be sick. The technical possibilities to realize them are manifold.

A particularly apt and impressive example is the transhumanist Martine Rothblatt, who was for a long time the highest-earning female CEO in the US. She was married to a woman as a man for a long time and has remained married to the same partner even after becoming a trans woman. She lives transhumanism not only in terms of her self-identity but has also realized it within her corporate concerns. Her company was responsible for transplanting a genetically modified pig heart into a human in early 2022. Traditional biological boundaries are being transcended by transhumanism. This concerns the boundaries between man and woman as well as the boundaries between human and non-human animals.

However, with the advancement of technical possibilities, many more ways of living diversity can be conceived. It is already possible today to realize children with three biological parents. This is done by producing a female egg cell, which is created from the cell nucleus of one woman and the rest of the cell of another woman. The artificially created cell then only needs to be fertilized and successfully implanted. The medical concern was initially to enable women with a mitochondrial disease to have biologically related children. In the meantime, the question has arisen as to whether the option of such children should not also be allowed to a polyamorous union of two women and a man without the presence of a disease. If this new family is then created, then they should also be given the option to marry, if this is in their interest. After all, it is almost a traditional family that would now consist of father, mother, mother, and biologically related child.

In the future, completely different options are conceivable, even if these are more likely to be realized in the far distant future, if at all. With the help of mind uploading, other body experiences could also be put into practice. There are only limits to one's own imagination here. What it is like to be a bat could then be experienced in practice, just like the possibility of being transferred from a male body into a female one and then back into a male one, but this time into a different one. The possibilities of different sensual experiences of the world could thus be increased enormously. Stimulating options could be inspired by Greek mythology. A particularly exciting example is that of the seer Teiresias, who was born male, then lived as a woman for a few years until she finally became a man again. He was confronted by Zeus and his wife with the question of whether sexual pleasure could be felt more strongly by men or women. Teiresias replied that women had a much stronger sense of sexual pleasure than men. It would be exciting to test this thesis in practice. For this, however, our technological possibilities would have to be much more advanced.

Whether, against the backdrop of such developments, there is not a need to move beyond the duality of the sexes and gender identities in general is quite obvious. What is decisive, however, is that Euro-Transhumanism clearly represents non-heteronormative ideas. How these are expressed in detail certainly differs among the various transhumanists. In any case, a decisive departure from Aristotelian notions of natural law can be observed.

Non-authoritarian

Transhumanism is also currently being intensively received among Russian intellectuals. Particularly noteworthy in this regard are the remarks of Dugin, who explicitly calls transhumanism, which he identifies with pluralism and many other 'evils', an idea of the devil.[20] The Patriarch of the Russian Orthodox Church, Kirill the 1st, has also frequently commented on transhumanism in a similar way.[21] As has already become clear within the considerations to be found here, Dugin's description of transhumanism is quite plausible in particular concerning Euro-Transhumanism. It is only his assessment that is not shared here. Rather than evil *per se*, I hope it has emerged that Euro-Transhumanism aims to promote the plurality of personal flourishing. Transhumanism is non-authoritarian and is also concerned with undermining problematic authoritarian structures.[22] With Dugin's Eurasian orthodoxy comes the authoritarian violence of truth in correspondence to reality. The fluid contingency, on the other hand, is concerned with reducing violence against persons and supporting the multiplicity of personal flourishing.

Conclusion

I hope that it has become clear that those who are interested in a new intellectual modesty, the best possible avoidance of violence against persons and the plurality of personal flourishing within a social liberal democracy should take a closer look at Euro-Transhumanism. There are still too many encrusted relics of past cultural structures that should be twisted, because with them a morally problematic violence against persons emanates by making clear: 'The way you live is wrong. You are sick. This is not how you should act.' By breaking down these barriers, new forms of experience can be realized. At the same time, the extension of the personal healthspan, a technically supported sustainability, and a more plausible relationship between human and non-human living beings are realized. To me, these revisions seem to go hand in hand with a higher quality of life, and I hope that I share these ideas with many other people. I am eager for the realization of our posthuman future to occur.

I also hope that I have provided the reader with some justification for why a philosophically informed version of transhumanist reflections can rest on central pillars of Nietzsche's thought. In general, it should be emphasized that a lot of reflection still needs to be done to develop a great plurality of proper philosophies of transhumanism. Transhumanism has most often been in the hands of scientists, entrepreneurs, and investors so far. By engaging transhumanist insights with philosophical reflections by Nietzsche, it is possible to realize complex, creative, and challenging philosophies of transhumanism. Thus, I have sketched a vision of Euro-Transhumanism (Schussler and Balistreri 2024).

Religion

Even though the word 'transhumanism' has been in use before, the concept 'transhumanism' was coined by Julian Huxley in 1951, whereby *foundational transhumanism* was established. Julian Huxley was a professor of zoology at King's College London, founding director of UNESCO and was involved in the conceptual background of the Universal Declaration of Human Rights. His grandfather was Darwin's bulldog Thomas Henry Huxley, his half-brother Andrew Fielding Huxley won the Nobel Prize for Medicine in 1963, and his brother Aldous Huxley wrote one of the most important utopian books of the 20th century with his novel *Brave New World*. Like George Orwell's novel *1984*, it was decisively inspired by the Russian novel *We* by Yevgeny Ivanovich Zamyatin. All three novels thematize political dystopias associated with the latest technologies. While Aldous Huxley implicitly warns against transhumanism in this way, his brother Julian Huxley initializes one of the decisive cultural currents of the present with transhumanism.

At the beginning of the 1990s, FM-2030, Natasha Vita-More and Max More established *original transhumanism* as an important cultural force in the public sphere. It differs significantly from *classic transhumanism*, which was developed by and in the circles of Nick Bostrom in the early 2000s. Its impact unfolded itself in particular by means of the Future of Humanity Institute at the University of Oxford, which he led from its foundation in 2005 until its closure in 2024 and which was financially supported by Elon Musk among others. Euro-Transhumanism, on the other hand, crystallized at the latest with the publication of Stefan Lorenz Sorgner's article 'Nietzsche, the Overhuman, and Transhumanism' in the *Journal of Evolution and Technology* in 2009, which unfolded an enormous amount of reactions, and connected transhumanism with reflections from continental philosophical traditions.

In contrast to the classic transhumanism of Elon Musk and his circle of friends, Euro-Transhumanism is decisively influenced by the continental European cultural tradition and distances itself decidedly from its libertarian basic attitude, which goes hand in hand with a dangerous political utopia and which assumes the coming about of a silicon-based further development of humans, according to which we will already exist as downloaded personalities on hard drives in a few decades and which highlights that humans will possibly be put into zoos by superintelligences. Euro-Transhumanism considers it more likely that we will first evolve as carbon-based beings, as 'we have always been cyborgs'.

The realization that we have always been cyborgs represents a radical break with the dualistic Western culture of the past 2,000 years, which has always understood humans as dualistic beings, that is, as entities with a material body and an immaterial rational soul, with which the unchangeable human nature, the divine spark, the image of God, free will, autonomy is identified.

A cyborg is an organism controlled by or upgraded with language, a *kybernetes*, that is, the ancient Greek for the helmsman, of the org, that is, the organism. The organic prerequisites for acquiring the ability to speak arose around 300,000 years ago as a result of a genetic mutation. From then on, it was possible to upgrade certain organisms through education so that they were able to communicate by means of language, develop further mental abilities and realize special personal characteristics. A complex interplay of both environmental and genetic changes leads to these specific characteristics.

The traditional narrative of the Christian West was that humans became rational by means of a divine spark. Since 1869, the Catholic Church has argued that a rational soul joins the human body at the moment of fertilization, which is why humans already possess dignity from this point onwards. The Euro-Transhumanist narrative that we have always been cyborgs tells an alternative story of how we came to reason, namely through the complex interplay of a genetic mutation and the parental and cultural upgrading that occurs in the context of education. This view is also widely shared in classic and original transhumanism.

In general, however, Euro-Transhumanism represents a cultural and philosophical alternative to the Anglo-American dominated narrative of classic transhumanism. The euro-transhumanist approach criticizes the libertarian, utopian, and naively hedonistic tendencies of classical transhumanism – epitomized by its focus on conquest, individual wealth, and superficial experiences – and proposes instead a model rooted in European intellectual and cultural traditions. Drawing on art, literature, philosophy, and music, it argues for a nuanced perspective that utilizes the wisdom of the past (for example, Heraclitus, Nietzsche) to grasp the complexity of the posthuman paradigm.

Utopias versus visions

In contrast to original transhumanists like Max More, classic transhumanists, such as Nick Bostrom, Elon Musk and Peter Thiel, are in favour of utopias. Euro-Transhumanism considers this basic attitude to be extremely dangerous. This assessment applies to all forms of utopias and not just transhumanist ones, that is, also to the utopia of a classless future society, the utopia of bodily resurrection at the end of days or the utopia that we will live as uploaded personalities in a digital society controlled by a benevolent superintelligence. The central danger of any utopia is to sacrifice the present in order to realize a perfect future social order that can never be implemented in practice. In utopias, the lives of people today are sacrificed to a future cloud cuckoo land, which is dangerous because it undermines the promotion of the quality of life of all people living today. Today's people are instrumentalized for the realization of a distant utopia.

Utopias have to do with ultimate justifications. Visions, on the other hand, are proposals that invite further reflection and discourse. It is this openness to alternatives and better scientific knowledge that characterizes Euro-Transhumanism. Utopias are opposed to such a basic attitude. All worldviews that affirm utopian goals are dangerous and should be rejected and fought against from a Euro-Transhumanist perspective. A non-utopian attitude goes hand in hand with the possibility of contingent visions, which can serve as a basis for further considerations. Euro-Transhumanist objectives are to be understood as visions and do not represent utopias.

Compared to the classic transhumanism of Silicon Valley, Euro-Transhumanism emphasizes the need for a new culture of transhumanism and not one in which the aim is to become a billionaire, live without meaningful ties and travel the world to create new experiences and Instagram stories and turn start-ups into unicorns in pop-up cities on Roatán or in Chiang Mai. While classic transhumanism focuses on a lifestyle characterized by techno, Thailand, tantra, and tattoos, Euro-Transhumanism focuses on the virtues of *grandezza*, *raffinatezza*, and *fiducia*. It emphasizes that we need art, literature, and music, as well as an engagement with Heraclitus, Spinoza, Goethe, Schopenhauer, Wagner, and Nietzsche in order to live a flourishing life. The future needs the past in order to transform our dualistic ontological binaries and to realize the posthuman paradigm shift.

As recently as 2001, Jürgen Habermas aptly pointed out that transhumanism had not yet had a broad intellectual impact. At the time, he characterized transhumanists as freaky intellectuals and self-stylized Nietzscheans with all-too-familiar motifs of a very German ideology. This assessment was no longer accurate only a short time later, which is why he categorized transhumanism as a cult-like activity in 2014, which is not a plausible assessment either, as it neither affirms rites, players, nor any other features which are central for cults.

By 2013 at the latest, however, transhumanism had made a decisive impact on the general public with the publication of Dan Brown's global bestseller *Inferno*. Since then, transhumanism and numerous phenomena associated with transhumanism have been represented in the media everywhere. At the Burning Man Festival, in Hollywood and Netflix blockbusters, and also as part of the Venice Biennale, transhumanism has been implicitly and explicitly addressed ever since.

Transhumanism developed a strong impact early on, particularly in films. Arthur C. Clarke, who studied physics and mathematics at King's College London, wrote the screenplay for the 1968 Stanley Kubrick film *2001: A Space Odyssey*, which is considered a classic film in the context of transhumanism. While this film aptly depicts a specific transhumanist utopia, the 1997 film *Gattaca* represents a technological dystopia in an enormously entertaining way, as it depicts a society in which the significance of a person depends on the quality of their genes and not on their membership of the species *Homo sapiens*. The 2014 film *Transcendence* starring Johnny Depp was a financial flop, but it aptly portrayed a utopia of classic transhumanism, namely the utopia of a digital human personality. It is therefore hardly surprising that Elon Musk makes an appearance in this film as Elon Musk. Elon Musk also makes a guest appearance in the extremely entertaining and hugely successful US sitcom *Big Bang Theory*. Transhumanist themes, motives, and reflections are also regularly discussed, portrayed, and debated here. Even fascinating transhumanist artworks are referenced in this series. The green fluorescent rabbit 'Alba' stands out in particular. It is one of the most remarkable bioart works by the artist Eduardo Kac, who also coined the term 'bioart' in 1997. What exactly can be described as bioart is still an open question. Patricia Piccinini's sculpture *Still Life with Stem Cells* from 2002 could just as easily be described as bioart as the famous modelled ear on the forearm of the performance artist Stelarc.

Even though the works of Kac, Piccinini, and Stelarc are often mentioned in the context of transhumanism, they have stronger intellectual links to Euro-Transhumanism than to classic transhumanism. This also applies to the artistic work of Sven Helbig, Joanna Grochowska, and Tatyana von Leys, who have a particularly close connection to Euro-Transhumanism. Sven Helbig's music drama *Vom Lärm der Welt* (On the Noise of the World) is one of the most fascinating operas of our time. Joanna Grochowska's sculptures deal explicitly with the intellectual creativity of continental European cultural history. Tatyana von Leys' paintings are deeply anchored in the complex web of posthuman reflections.

It can be assumed that the age of transhumanism has only just begun with these developments. The effectiveness of technologies is constantly intensifying. CRISPR-Cas9 was developed in 2012. ChatGPT was presented to the public in 2022. Musk's Neuralink project 'Blindsight' enabled some monkeys to see using brain–computer interfaces in 2024.[1] The possibilities of the latest technologies are enormous.

Love, sex, gender, religion

When it comes to religion and politics, many philosophical issues are related to how sex and gender are conceptualized. In Western history, the

Aristotelian natural law tradition has played a significant cultural role for promoting a discriminatory binary way of thinking rooted in a categorically dualist ontology. The eyes are there to see. The nose is there to smell. The genitals are there for reproductive purposes. If they are not used for this purpose, then they are misused. They are used in an unnatural manner. They are used in a manner that undermines personal flourishing. This has been the dominant understanding in a certain natural law tradition, which has also been integrated into Catholicism. This attitude leads to heteronormativity as well as to sex only being seen as natural if it occurs without contraceptive and such that it can lead to a pregnancy, which excludes anal or oral sex, homosexual sex, as well as masturbation. It represents one of the morally problematic discriminatory aspects which goes along with humanism.

The posthuman–paradigm shift aims to move away from categorically dualist ontologies, which have such discriminatory implications. The great variety of beyond humanism movements suggest different options for avoiding discriminations like sexism, racism, speciesism, or heteronormativity. This applies to critical posthumanism as well as to most transhumanist approaches, like original transhumanism which is connected with Natasha Vita-More and Max More as well as to Euro-Transhumanism. Some of the most poignant transhumanist reflections on this issue have been presented by Martine Rothblatt in the monograph *From Transgender to Transhuman: A Manifesto on the Freedom of Form* (2011).

Unfortunately, there are some recent cultural developments which provide further intellectual nourishment for another widely shared prejudice, namely that transhumanism is merely a humanism on steroids, which aims for a Renaissance ideal of perfection. Such prejudices get further support from reflections presented by classic transhumanism, which can be associated with selected members of the Future of Humanity Institute which existed at the University of Oxford in between 2005 and 2024. This institute received a considerable amount of financial help from Elon Musk, who also publicly defended specific reflections put forward by the institute's former director. All of this strengthens the insight that Elon Musk's acts and reflections can only be understood properly with an awareness of classic transhumanism, whose relevance today can hardly be underestimated given recent political developments.

Elon Musk is the richest person in the world, and he has become massively influential in recent times both in the US as well as in other parts of the world. He affirms a central insight which is representative of classic transhumanism, namely that it is highly likely that we are already living in a computer simulation. Without a comprehensive understanding of classic transhumanism, it is difficult to fully grasp many of his decisions and utterances. Classic transhumanism also represents the intellectual grounding of Peter Thiel and James O'Neill. Peter Thiel founded PayPal together with Elon Musk and created the Thiel Fellowship together with James O'Neill. In 2014, Vitalik Buterin, the future inventor of the cryptocurrency Ethereum, was awarded this fellowship. In 2021, *Time* magazine named Vitalik Buterin one of the 100 most influential people in the world. In 2018, Buterin donated more than $2 million to the SENS Research Foundation. The aim of this foundation is to increase the human healthspan by aiming to combat the disease of ageing. This is a central transhumanist objective, which is accompanied by either the utopia of immortality or at least the vision of an indefinite extension of the healthspan. From 2019 to 2021, the aforementioned James O'Neill was CEO of the SENS Research Foundation. Following his re-election as President of the United States of America in 2024, Donald Trump announced his intention to nominate James O'Neill for the post of US Deputy Secretary of Health and Human Services. He was sworn in as Deputy Secretary of Health and Human Services (HHS) of the United States by Secretary Kennedy on 9 June 2025.

Given these developments the intricate entanglement of politics and classic transhumanism can hardly be underestimated. This has significant practical consequences, in particular if certain decisions and utterances from Musk and his friends are being considered. According to a BBC report from 2024:

> Elon Musk has said he will move the headquarters of two of his most high-profile companies, rocket firm SpaceX and social media platform X, from California to Texas. He said the move was due to recent laws passed by the state – in particular a new law which prevents schools from making rules requiring staff tell anyone, including parents, information about a child's gender identity.[2]

After having become president again in 2025, Trump immediately made some decisions which express his sympathies towards a binary way of thinking, as he:

> issued an executive order rescinding two orders that Joe Biden signed on his first day in office four years ago, one advancing racial equity for under-served communities and another combating discrimination based on gender identity or sexual orientation. One executive order, called 'Defending Women from Gender Ideology Extremism and Restoring Biological Truth to the Federal Government', instructs the federal government to remove 'all radical gender ideology guidance, communication, policies, and forms'.[3]

While the majority of transhumanist approaches support LGBTQIA+ rights, this does not necessarily apply to classic transhumanism, which is politically extremely influential. From the Russian orthodox perspective, which is presented by Dugin and his friends, like the Patriarch of Moscow Cyril I, who also participated in many anti-transhumanist events, transhumanism can be identified with the LGBTQIA+ movement as well as with many other 'evils'.[4] Interestingly, a strong criticism of transhumanism also comes from another ally of Trump, namely his former advisor Steve Bannon, who wrote the preface of the anti-transhumanist book *Dark Aeon*, and who recently highlighted that 'Elon Musk Is an "Evil Person"'.[5] The former White House chief strategist also agreed with Fukuyama's assessment from 2004 that transhumanism is the most dangerous idea in the world. However, Bannon also emphasized that the practice is even more dangerous than the theory, associating transhumanism with an aberration of humanity that was solely in the interests of the global world elite, with this criticism also containing anti-Semitic connotations.

It is also noteworthy in this context that Bannon and Dugin met personally in Rome in 2018 to discuss many philosophical issues related to traditionalism.[6] This outlook is also being shared by Olavo de Carvalho, who is 'a far-right Brazilian pundit and self-proclaimed philosopher who became the political guru of President Jair Bolsonaro'.[7]

> At its core, traditionalism rejects modernity and its ideals: faith in the ability of human ingenuity to advance living standards and justice; an emphasis on the management of the economy; the coveting of individual liberty; the existence of universal truths equally valid for, and thereby equalising of, all. Repudiating the Enlightenment, traditionalists instead celebrate what they regard as timeless values. They honour precedence rather than progress, emphasize the spiritual over the material, and advocate surrender to the fundamental disparities – as opposed to equality – between humans and human destinies.[8]

This additional aspect of the current political landscape adds further spice within the camp of Trump supporters. Given that there is a clear pro classic transhumanism stance (Musk) as well as a decisive anti-transhumanism attitude (Bannon) within the group of Trump supporters already seems to prefigure a strong conflict within this camp. What is intellectually interesting given the recent political developments concerning gender issues is that it further supports the insight, widely shared among critical posthumanists, that a certain version of transhumanism, namely classic transhumanism, can be conceptualized as hyper-humanism. On the other hand, it needs to be kept in mind that original transhumanism, Euro-Transhumanism, and in particular influential transhumanists like Martine Rothblatt take a diametrically opposed stance concerning such issues, as they strongly support the LGBTQIA+ community. It is this understanding of transhumanism which is seen as characteristic of transhumanism from the Russian orthodox perspective of Dugin and his friends.

It is interesting to note how fundamental the concepts of sex and gender are for the various, cultural, political, as well as a philosophical debates on transhumanism, in particular when it comes to questions concerning religion.

Religion, perspectivism, Dukkha, and transhumanism

According to Dugin, transhumanism is the idea of the devil. However, what do transhumanists say about this judgement? As a matter of fact, it needs to be stressed that there are Catholic, Jewish, Muslim, Evangelical, Lutheran, Buddhist, as well as Mormon transhumanists. On the other hand, it also needs to be highlighted that most transhumanists seem to be atheists, naturalists, and sceptics.[9] The decisive question is how the relationship between religion and transhumanism can meaningfully be analysed further from a philosophical perspective, whereby, I will provide some further reflections from the perspective of Euro-Transhumanism, even though they might also apply to other transhumanist approaches which share the understanding that human beings are fully integrated into the world of permanent becoming at all times in all respects, or a Buddhist ontology of becoming and Dukkha. Given such an outlook, it is no longer plausible to uphold a traditional epistemology which claims the possibility of true judgements such that truths as correspondence to the world can be grasped. If this is the case, then, perspectivism remains as the most plausible epistemological approach.

What does perspectivism stand for, mean, and imply? All philosophical perspectives are interpretations. This means that they can be wrong, but also that they do not have to be wrong. However, rightness and wrongness are not decisive criteria. Instead, what matters are 'constructive interferences' with one's core drives, needs, and longings. Epistemologically, there are no better or worse perspectives. However, with respect to one's drives, better or worse perspectives exist, some resonate more strongly and positively than others. There are constructive as well as destructive interferences, but constructive interferences are meaningful for one's own existence-

These reflections lead to the realization that all philosophical judgements are caught in a hermeneutic circle. All philosophical perspectives are interpretations. This also applies to perspectivism itself. However, criteria for better and worse and for decisions are needed for making decisions. Where do these criteria come from? They do not come from anywhere outside of yourself. These criteria exist in your drives. What is challenging is to distinguish one's core drives from others which have been installed within you by external forces like social media platforms. Being able to distinguish between these two sources is a challenging issue.

These reflections also lead to aletheic nihilism, as there no longer seems to be the option of making any truth judgements, which correspond to the world. It is this insight, this Nietzschean desert, which most thinkers from the Western philosophical tradition regard as a challenge, which needs to be overcome. To embrace the desert, to see the desert as a gain rather than as a challenge which needs to be overcome is an insight which my teacher, Gianni Vattimo, stresses. I regard his reflections on this specific issue as highly plausible. Vattimo's reflections explain why nihilism is an enormous gain which needs to be cherished immensely, and why any attempt to move beyond nihilism has highly problematic paternalistic, violent, and potentially totalitarian implications.

I fully grasped the enormous impact of these insights after personally having made his acquaintance in November 2000 during an international conference organized by the Nietzsche Forum in Munich, Germany, entitled 'Die Auflösung des abendländischen Subjekts und das Schicksal Europas' (The dissolution of the Western subject and the fate of Europe). Nihilism and the crisis of values have been central philosophical challenges in the past 200 years. It has immediately become clear that Vattimo lives his philosophical reflections and is personally engaged with these reflections. He does not present a distanced armchair philosophy, which he developed by means of a 9 to 5 activity which is primarily connected to an intellectual engagement with written reflections, but his thinking is related to a hermeneutic lifeworldly embeddedness in which the philosophical reflections are put into practice, and revised again, if the thoughts do not match with experiences.

From a cultural perspective the traditional Catholic outlook was shattered after it was confronted with criticisms of the various ethical, epistemological, and ontological absolutes during the Enlightenment. In addition, the tensions between the normative Christian duties and psychophysiological demands for living a fulfilled life have become more evident within liberal political structures. This led to the intellectual need of a reorientation, a period of chaos, intellectual uncertainty, and lived nihilism. This is a short and abstract description of cultural processes which have taken place during the past 300 years. At the same, this process is often shared by humans during their teenage years after they have experienced a rigorous religious education. Culturally, it is clear that with Darwin, Nietzsche, and Freud the certainties of Catholicism have been shattered, and God has died, as Nietzsche stresses. God had been alive and active for nearly 2,000 years. The Christian truth has slowly lost its cultural impact during the past 300 years. Thus, truth vanished, and hermeneutics has steadily become more and more important.

From a personal perspective, Vattimo noted that the older he got, the more important Catholicism became for him again, too. He also realized that hermeneutics and weak thinking do not justify a dogmatic affirmation of atheism. Perspectivism as well as weak thinking do not justify the dogmatic claim that God does not exist. On the contrary, perspectivism is an epistemological prerequisite for allowing an affirmation of Catholicism again. To claim that atheism is true is as dogmatic as the claim that Catholicism is true. None of these truth claims can be justified on the basis of perspectivism. However, Vattimo believes to believe in the Christian God. It is this belief which is not undermined by perspectivism. To believe in something implies a certain sort of doubt. Knowing something means that it is no longer necessary to believe and Vattimo believes in God. However, this type of Catholicism no longer is a dogmatic, essentialist, and absolutist understanding of Catholicism. Perspectivism demands a weak Catholicism, a Catholicism, which does not and cannot demand that it is the ultimate absolute truth which is universally valid. Hence, Vattimo's weak Catholicism is not a Catholicism which is in tune with the Aristotelian–Thomistic natural law tradition, which has been so strong in the Catholic church. Hence, it demands a hermeneutic reinterpretation of the virtue of love, the central Christian virtue, as St Paul clearly stresses: love, hope, and faith are important, but love matters most.

Given the relevance of love, Augustine already upholds: love and do whatever you wish to do. Augustine and the natural law tradition have a clear understanding of what love means. Each one of us ought to live such that our natural *telos* or aim gets realized, and I have already explained the heteronormative implications of such an understanding of love. This has been the dominant attitude towards love and sexuality for many centuries in the Catholic church. Vattimo's weak thinking proposes a different understanding of love. He also affirms to love and do whatever you want to do. However, his interpretation of love differs significantly from Augustine's, which has important implications both for individual morality as well as for cultural processes.

Given the dominance of the natural law tradition in the Catholic church, the Enlightenment has long been seen as a time which goes along with a moving away from Christianity, as politically liberalism, plurality, and the ethics of autonomy have steadily become more important in the past 500 years. However, Vattimo sees the ethics of autonomy, if it gets lived in connection with the virtue of love, as in tune with the preachings of Jesus Christ in the New Testament. Love and do whatever you want as long as you do not harm another person but rather take a supporting attitude towards others. If you live in a heterosexual, homosexual, or polyamorous relationship or if you love someone who is 40 years your senior, it is all fine, if this mutually resonating dedication occurs on the basis of the virtue of love. This is the paradigm shifting insight which goes along with Vattimo's weak Catholicism.

The punishing God of the Old Testament represents the natural law understanding of love. However, Jesus Christ in the New Testament replaced this understanding with a weakened understanding of love, in the same way as the entire history of the world can be seen as a weakening process, a process of kenosis, according to Vattimo. God creating the world is a weakening process. God becoming a human being with the birth of Jesus Christ is a weakening process. The rigid laws of the Old Testament being replaced by a weak understanding of love which has slowly gained momentum during the Enlightenment is a weakening process. Thus, the violent demands of the sacred dissolve, which further promotes a weakened understanding of the virtue of love (Vattimo 2003)

Catholicism, love, and church services are not seen as absolute truths by Vattimo, but they are seen as contingent nodal point which are being affirmed by him. He recognizes their contingency but embraces them both personally as well as culturally. On this basis, a cultural strengthening of Catholicism can become possible again, too, even though we still live in the desert.

It is this type of logic which is fully in tune with Euro-Transhumanism, and any version of transhumanism resting on an ontology of becoming, because on such a basis, perspectivism becomes plausible as epistemology. This epistemic attitude is a decisive reason, why any strong rejection of monotheistic religions is no longer an option, and affirming a belief in such a religion becomes a possibility again. However, at the same time, it has implications concerning the moral demands which go along with such an affirmation, as any dogmatic, essentialist, and universal normative ethical claims can no longer be upheld either. Hence, the following intellectual demands go along with embracing the epistemology of perspectivism: (1) an openness towards other cultures; (2) respect for the beliefs of another person; and (3) a certain modesty concerning one's worldview due to the realization of the contingency of one's own values, norms, and beliefs.

Conclusion

In this section, I stressed the relevance of distinguishing different versions of transhumanism, highlighting the political entanglements of the various options. This groundwork provides the basis for arguing that there does not have to be a conflict between the affirmation of transhumanism and the belief in a monotheistic religion, given that the previously mentioned three intellectual demands concerning one's own thinking and acting get considered, which have to do with a modesty concerning one's own self-understanding as well as respect concerning the worldview of others.

Politics

Liberal democracies are an amazingly outstanding achievement, as they are based on the understanding that each person subscribes to a distinctive concept of the good life and that persons should have the right to live following their understanding of what it means to live a good life.[1] The cultural relevance of this normative political foundation can hardly be underestimated. This is my philosophical stance, even if I acknowledge that I have no ultimate philosophical foundation for making this claim apart from the judgement that I cherish living in political circumstances in which a person is not forced to use totalitarian measures to stick to a strong concept of the good life. I am glad that this political attitude is widely shared in our times. However, it must be noted that several political forces today, which are dominant both in politics and academia, argue for a different type of political organization.

Critics of liberal democracies

The central distinction in this article is between thinkers who affirm some form of liberal democracy and those who do not. The website Our World in Data stresses that 'the world has become much more democratic over the last two centuries'.[2] This does not mean that the development towards more democratic countries is necessary, nor that any democratic government is a liberal democratic one. There is, furthermore, a wide range of scholars in humanities departments across universities who are highly critical of liberal democracies. Moreover, it must be acknowledged that several countries wielding significant (economic) power today still have authoritarian regimes or have implemented measures to reestablish more authoritarian structures.

A powerful opposition to liberal democracies comes from fundamentalist religious quarters. Many critics of liberal democracies are traditionally religious, political philosophers who affirm a rigid subject–object distinction. Dugin and the patriarch of Moscow Kyrill represent two intellectuals in this category. Dugin argues for the dominance of traditional orthodox Christianity in Eurasian countries. He regards transhumanism as the 'idea of the devil'[3] representing all the evils of the world, for example, liberalism, democracy, and plurality, which targets, in particular, the LGBTQIA+ community.

Besides religious conservatives who criticize liberal democracies, there is an excellent number of leftist positions arguing against the desirability of liberal democracies. Most of these are highly critical of the subject–object distinction. Postmodern communists, like Agamben, and a great variety of posthumanisms belong to this spectrum. Biopolitics is their crucial term. Agamben, who is widely known for his criticism of biopolitics, has used this term to attack enhancement. His positions are so scathing that even the French philosopher Jean-Luc Nancy, his friend, warned us of the dangers of Agamben's suggestions: 'if he himself had followed Agamben's medical advice discouraging a heart transplant that saved his life, that he would be dead'.[4] Bratton explains the special relationship between Nancy and Agamben. Furthermore, Bratton makes the following perceptive comment on Agamben's reflections on mandatory COVID-19 vaccinations during the pandemic in which Agamben compares state-mandated vaccination to the political structures of fascist Germany: these biopolitical regulations are supposed to have had totalitarian fascist implications and to have divided citizens into Aryans and sub-humans, or 'Second class citizens'; he connects the fates of those who refuse vaccination to that of Jews under fascism and concludes, 'The "green card" (Italy's vaccine pass) constitutes those who do not have it in bearers of a virtual yellow star'. After picking up my jaw, I cannot help but compare Agamben's analysis to that of QAnon-influenced US congressperson Marjorie Taylor Greene, who beat him to the punch when she tweeted, 'Vaccinated employees get a vaccination logo just like the Nazi's forced Jewish people to wear a gold star'.[5]

However, not only religious fundamentalists and some postmodern communists present highly dangerous totalitarian visions. The same applies to some posthumanists, namely several radical posthumanists, as I call them. Radical and critical posthumanists need to be distinguished. Braidotti is the most notable critical posthumanist who falls into the category of critical posthumanists with solid reservations concerning liberal democracies. Radical posthumanists represent views that are anti-natalist and are generally interested in promoting a future in which rich, white, heterosexual men – the villains of history – first go extinct, followed by the death of humans in general, which they regard as essential for the promotion of the planet's health. According to them, human beings are parasites on the earth. However, it is in the planet's interest that these colonial structures vanish, which can be realized best without humans. Radical posthumanists cannot explain, however, why rats or oak trees should matter more than human beings.[6] They also cannot make philosophical sense of the expression 'health of the planet'. Health is a concept that is notoriously difficult to define. However, if it can be employed meaningfully, it can only be used for organisms. The challenge is that the entire earth is not a unified organism; it is incapable of self-movement and does not possess a specific unified metabolism. Consequently, health cannot be meaningfully employed to the entire earth. Radical posthumanism is not an approach that deserves further philosophical consideration.

Radical posthumanists, postmodern communists like Agamben, and religious conservatives present political utopias with strong totalitarian implications. After the experiences of the 20th-century totalitarian regimes, it should have become clear that we must fight against philosophical and political ideas with such dangerous totalitarian implications (Sorgner 2022a).

The situation is slightly different concerning critical posthumanists, who argue against the plausibility of liberal democracies and who stress the 'posthuman relational subjectivity and the politics of affirmation' as Braidotti does.[7] Many critical posthumanists affirm a relational subjectivity and a relationalism that they apply to democracies, ethics, and politics. A liberal understanding, on the other hand, rests upon personal rights, political subjectivity, and a concept of personhood. Here, the implications of relationalism must be further explored. They are most clearly visible in the political practices in contemporary China, as China has integrated significant structures from the relationalism of its religious and cultural history into its contemporary political systems.

In China, the following occurred recently.[8] A Chinese boy applied to a university and had to take entry tests, which he passed. However, his father had yet to meet payments on his loan, resulting in a low social credit score. Entities do not count as individuals responsible for their actions in a relational, political structure. Still, a relational subject's duties and obligations depend on the relational network in which they are embedded. Due to the relevance of the relational embeddedness of the young applicant and the fact that his dad had yet to pay back the loan in time, he was rejected from the university, even though he had fulfilled all the admission requirements for entering the university. In a relational political system, a person's values and norms do not depend on their qualities and capacities but on the relational network in which the entity is embedded.

A relational, political system that promotes diversity would prefer an outcome in which an oak tree, a rat, and a human being are saved over a scenario in which three humans are held, with the rat and the oak tree dead. Here, the challenging implications of a relational political structure emerge. It thus becomes clear how these values do not accord with the notion of liberal democracies, which are grounded in personal rights. However, to save a rat and an oak tree instead of two human beings or to forbid a young boy from entering university even though he has fulfilled all admission requirements is unjust, as it has totalitarian implications that unjustly sacrifice human ambition and flourishing. Such structures are violent, paternalistic, and oppressive against persons who have not done anything blameworthy.

They are sacrificed by virtue of their position within a network, leading to their unfair punishment.

Even though I affirm the critical posthumanists' assertion that the subject–object distinction is ontologically obsolete, I disagree that this ontological insight must have consequences on the political relevance of subjectivity. By claiming that a relational ontology demands a relational ethics implies committing the is–ought fallacy. What ought to be done cannot logically be derived from what is ontologically the case. While a relational ontology of permanent becoming in all respects at all times is highly plausible (Sorgner 2022b), a relational ethics is highly dangerous, and not necessary at all. In an ethical and political context, the concept of the person and the subject is pragmatically meaningful. The concept of political subjects is an extremely helpful fiction for a well-organized political system. Upholding political subjectivity might not be an ideal solution to political challenges, but it is an as-good-as-it-gets solution (Sorgner 2022a). This is the reason I hold the concept of the political subject to be significant. All persons should be counted as political subjects. The tricky question is who should count as a person. Elsewhere, I stressed that the concept of a person should not be identical in scope to that of human beings but rather open to non-human animals as well as embodied AI. Liberal democracies where the rights of all persons are considered are an enormous achievement. However, there is a wide range of liberal democracies.

Defenders of liberal democracies

Not all critical posthumanists are critics of liberal democracies, even though they are critical of the subject-object distinction. There are ways of justifying different kinds of subjectivity, for example, subjectivity as a constitution, as a metaphor, or as a fiction. Still, it must be noted that the majority of critical posthumanists highlight the relevance of alternative, grassroots, or non-dualistic versions of democracies. Most critical posthumanists affirm a relational ethics as well as politics. This also implies the rejection of a rights-based account of political governance as well as the rejection of a notion of dignity or personhood, which implies the relationship of equality among all the bearers of this quality. Critical posthumanists highlight that by applying notions like personhood, morally problematic, paternalistic, and oppressive hierarchies are brought about. It is a central goal of critical posthumanists to move beyond colonial, oppressive, and universalist hierarchies. Yet, the previous examples and reflections demonstrate that relational politics has highly violent, dangerous, and oppressive totalitarian implications. Liberal democracies might not be without moral challenges, but I regard them as an as-good-as-it-gets solution for a well-functioning political system.

What does this mean concerning a specific situational challenge? With the foundation, I suggest, even neo-colonial worries can be considered. There are still worldviews which practically demand female genital mutilation or the burning of 8-year-old girls as witches. An ontology of permanent becoming epistemologically renders perspectivism plausible, that is, which means that all philosophical perspectives are interpretations, whereby an interpretation is a judgement that can be false, but which does not have to be false. Here, it needs to be noted that philosophical judgements clearly need to be distinguished from empirical and scientific judgements, which can be falsified. Philosophical judgements deal with foundational explanations or in Goethe's words: '*was die Welt im Innersten zusammenhält*' (my translation: 'what holds the world together at its core'). Given this perspectivism, it follows that ontologically the previously mentioned worldviews with all their brutal implications do not have to be false. Hence, I as a white male, do not take a neo-colonial stance when I criticize them, as I acknowledge, at the same time, that ontologically they do not have to be false. However, if possible, I would do my best to stop female genital mutilation and the burning of 8-year-old girls as witches from occurring, as here direct harm against an innocent person is being done, which I regard as morally wrong. However, I acknowledge that I do not have a foundational basis for justifying the epistemological superiority of this moral principle. This also implies that I do not have an ultimate foundation for justifying the epistemological superiority of liberal democracies. However, I present various narratives for revealing why I prefer living in a liberal democracy to living in an authoritarian political system.

In contrast to critical posthumanists, all leading transhumanists belong to the liberal democratic spectrum (Sorgner 2020). It is essential to distinguish between post- and transhumanists (Ranisch and Sorgner 2014), even though many scholars use posthumanism as the umbrella and regard critical posthumanism and transhumanism as versions of posthumanism. This does not correspond to the use of the terms by leading transhumanists. A further central reason for clearly distinguishing transhumanism and posthumanism is that their cultural pedigrees differ significantly. Posthumanism has its roots in French postmodern philosophies, whereby the approaches by Deleuze and Foucault are particularly important. It is metaphorical in its rhetoric. It is solid in literary, cultural, and media theory. It embraces a dialectic manner of thinking. Their central goal is to move beyond the categorical ontological binaries usually associated with rigid opposing evaluations. Their concern for justice leads them to affirm the standpoint of the oppressed to establish a new order. Posthumans, in the posthumanist sense, are a non-dualistic reinterpretation of what used to be an essentialist understanding of human nature identified with an immaterial, rational, and autonomous soul and a material body in our Judeo-Christian past.

Conversely, transhumanism can be traced back to the Anglo-American tradition of evolutionary theory, in which the Huxley family played a central role. Julian Huxley coined the term in 1951. It can be associated with a scientific, linear, and rights-based philosophical approach. All classic transhumanists are utilitarians. This approach is dominant in the Anglo-American world. However, this judgement does not apply to original transhumanism. The central concern of transhumanism is to transcend our limitations to increase the likelihood of persons living good lives. Thus, the possibility of the coming about of posthumans also increases. Posthumans, in the transhumanist sense, further develop the human species. In contrast, posthumans, in the posthumanist sense, represent a non-dualist cultural redefinition of what it means to be a *Homo sapiens*.

Even though all leading transhumanists belong to the liberal spectrum, there is a wide range of different takes on which type of liberal democracy is affirmed. Technoprogressive transhumanists, like Euro-Transhumanists, represent the social, liberal, democratic end of the political spectrum, and libertarian transhumanists, like classic transhumanists, are mainly concerned with an affirmation of negative freedom, the absence of constraint, and thus, it represents the closest we can get to a meaningful understanding of anarchy. Technoprogressivism has been summarized by Hughes as follows:

> Technoprogressives, like social democrats in general, believe democracy requires strong guarantees of civil liberties and minority rights, a relatively egalitarian distribution of wealth and power, a strong state accountable and transparent to its citizens, and a process for deliberation and decision-making open to all competent persons. We are critical of the way that unequal power in society, based on gender, race, class, religion and so on, distorts the democratic process, and we insist that state agencies are necessary and that they should be independent of corporate and sectoral influences.[9]

Some technoprogressive transhumanists' and critical posthumanists' political guidelines might be similar. However, it should be noted that critical posthumanists like Katherine Hayles attempt to move beyond the categorical subject–object distinction, whereas this is only the case for some transhumanists. As a Euro-Transhumanist (Sorgner 2022c; 2009), I share her doubts about the ontological plausibility of categorical subject-object distinctions. However, this does mean that a rights-based approach must be undermined, as I regard the autonomous subject as a pragmatically helpful fiction (Sorgner 2022a). It does not exist, but politically, affirming the fiction of independent subjects is advisable. This anthropology positions the anthropology of Euro-Transhumanism quasi-in-between critical posthumanists and classic transhumanists, which is also the reason why this approach can be called metahumanism. *Meta* is ancient Greek and means in between as well as after. My version of metahumanism moves beyond a traditional binary version of humanism but also lies between critical posthumanism and classic transhumanism (Sorgner 2021).

Transhumanists James Hughes (2010) and Steven Fuller (2011; 2012; Fuller and Lipinska 2014) subscribe to the subject–object distinction. The same applies to a classic transhumanist like Nick Bostrom. In contrast to the former two thinkers, Bostrom affirm a utilitarian libertarian version of transhumanism, that is, classic transhumanism.[10] Concerning ethical judgements, Bostrom is close to transhumanists like Elon Musk and Peter Thiel, who are most strongly represented in the media. They represent the radical libertarian end of the liberal democratic spectrum. Even though it is fashionable among critical posthumanists, bioconservatives, traditional Marxists, and many religious leaders to identify transhumanists with blood-sucking vampires in Silicon Valley, this attitude does not apply to all transhumanists. Most transhumanists working in academia align themselves more with technoprogressivism, affirming a social liberal version of democracy.

Besides libertarian transhumanists, some libertarians do not self-identify as transhumanists. John Harris (2011) and Julian Savulescu (2001) are two particularly noteworthy thinkers regarding the politics of enhancement. Both are utilitarian libertarians, but neither is a transhumanist, even though they affirm positions on the ethics and politics of enhancement that might be significantly stronger than the positions of many transhumanists. The decisive difference between bioliberals, such as the political libertarians Savulescu and Harris, and transhumanists is the issue concerning the desirability of the coming about of the posthuman. Transhumanists regard the coming about of the posthuman as a desirable goal, whereas bioliberals do not affirm this judgement.

Many traditional defenders of liberal democracies, like Sandel (2007), Fukuyama (2004), and Habermas (2014b), can be classified as bioconservatives. Conservative religious philosophers, many postmodern communists, and most critical posthumanists also need to be classified as bioconservatives. Religious philosophers worry that humans play God by drawing on enhancement technologies. This attitude should represent a hybris that would lead to human extinction. Communists are particularly worried about the issue of people experiencing poverty exploited by owners of capital. Critical posthumanists often share the former concerns. In addition, they are worried that evil technologies will undermine human nature, that new forms of colonialism will be promoted, and that speciesism will be further advanced as they identify transhumanists with rich, white, male, heterosexual scientists who are merely interested in improving the lives of those near and dear to them, and who wish to become immortal by downloading their personalities onto hard drives.

Defenders of liberal democracies who are not religious philosophers, communists, or critical posthumanists, on the other hand, are primarily worried about equality when it comes to enhancement technologies. This moral issue is particularly stressed by Fukuyama, Sandel, and Habermas. It is also why all of them can be classified as bioconservatives. They all affirm a rigid subject–object distinction and highlight that our understanding of human subjects is the foundation of liberal democracies. All human beings are subjects (bearers of dignity), and all subjects (bearers of dignity) deserve the same kind of normative respect, which means that all of them must be treated equally well. Their descriptions are often such that the concept of human dignity or that of a human subject has both a descriptive and a normative dimension. Human beings are subjects, which descriptively makes them equal (Sorgner 2010c).

Consequently, all human beings deserve to be treated equally, too. They normatively deserve the same level of moral respect. A general worry of the previously mentioned liberal ethicists is that the use of enhancement technologies can undermine our descriptive equality, which could also damage the normative equality of all human beings, the foundation of our humanist world. This is the reason why the conservative political scientist Fukuyama regards transhumanism as the most dangerous idea in the world. In 2023, Trump's former adviser Steve Bannon supported Fukuyama's judgement, highlighting that this 'aberration of humanity' is in the interest of the 'world's elite corporate overlords' as 'the global institutions of finance, Wall Street, and Davos are' behind this endeavour (Bannon 2023). Thereby, he goes beyond the criticisms mentioned earlier. The terms 'globalism', 'cosmopolitanism', and 'world's elite corporate overlords' have often been understood as a code for 'Jews', so it seems that these comments add an anti-Semitic connotation to the criticism of transhumanism.

Enhancement and equality

Can enhancement technologies undermine equality? What kind of equality are we talking about, descriptive or normative equality? Can it also be the case that these enhancement technologies can be used to promote equality? All political subjects are equal and should be treated equally well. They all possess equal rights and deserve to be treated according to these rights. All these phrases seem clear, true, and even self-evident to many intellectuals today. However, to what extent are all political subjects equal? Who is a political subject? Do we talk about all human beings or only some? Can there be a non-human personhood? Can personhood apply to embodied AI or also to non-human animals? Does a fertilized egg possess personhood? When does personhood begin? With fertilization, two weeks after fertilization, when we are an embryo, when we are a foetus, when we are born, when we can recognize ourselves in the mirror, when we are autonomous, when we are competent judges, when our brain development is completed, when we have a sufficient amount of experiences? In Germany, you can only become president after having celebrated your 40th birthday. When do we lose the status of being a subject at the end of life? If passing the mirror test is relevant, should the subject status not also apply to dolphins, chimpanzees, and magpies?

Both the beginning and our end as political subjects are challenging issues. It is even more complicated when we consider the latest scientific advances. In 2023, scientists created synthetic embryos.[11] They did not undergo fertilization, but were compiled, created, and put together such that they came about and looked like an embryo on day six after fertilization. Yet these synthetic embryos did not experience fertilization or the first five days after fertilization. Should they be counted as political subjects? A subject that has the right to vote? However, the right to vote is usually only given to humans in their late teenage years, when they have reached a certain maturity. Should you be able to lose your rights again?

Traditionally, only rich white men had the right to vote. Only they were counted as subjects. The same happened with some enslaved people in England, who entered court first as objects and left it as subjects. Here, a transubstantiation occurred as a consequence of the judge's decision that someone should be treated as a person.

Who counts as a subject, a person, or a being with dignity is an enormously challenging issue. Bioconservative liberals worry that enhancement technologies undermine our chance at equality. Enhancement technologies only alter the body. The soul is traditionally considered an immaterial entity; enhancement technologies cannot modify such entities. Why is Fukuyama worried that enhancement technologies undermine our status as equal subjects? Is there any equality when it comes to our bodies? About 99.6 per cent of all genes are identical in two different human beings. At the same time, one must recall that 60 per cent of all genes are similar to a banana and 98.8 per cent to a chimpanzee (depending on the study). It is a central worry of bioconservative liberals that by altering our genes, we undermine the bond of what we identify as human beings. Why do we care about 99.6 per cent of all genes? Why should we not expand the concept of subjectivity to chimpanzees, with whom we share 98.8 per cent of our genes? It seems a contingent decision. It could even be argued that two human beings have nothing in common.

Regarding their spatiotemporal data, no two human beings can have identical qualities. Even when it comes to genes, it could be stressed that genes can be turned on and off due to epigenetic effects. In addition, our bodies are constantly interacting with the environment. Hence, our bodies are always in the process of change. If we apply the most robust Leibnizian concept of identity, including spatiotemporal qualities, we are not even identical with ourselves over time.

The question of identity over time concerning human bodies is enormously challenging. To trace human equality to a bodily quality, which usually implies a type of identity, is a highly problematic enterprise from a philosophical perspective. There is no human physical identity over any period of time. Why, then, would normative equality between subjects get lost if enhancement technologies are employed? Human enhancement technologies alter the body and cannot change any immaterial entities. There is no identity in the most robust Leibnizian sense between two human beings over any period regarding bodily qualities. Whether human bodies get altered using human enhancement technologies or not should not matter at all concerning the moral status of these human beings, as it is merely the body which can be altered, and there is no bodily identity between any two human beings when we employ the notion of identity in the strongest Leibnizian sense.

Maybe, these bioconservative liberals have a different notion of equality in mind. It seems that the worry that our equality will be undermined is evoked, particularly when the issue of gene technologies is addressed. If we managed to introduce selected genes of sharks, whales, and bats into the human genome, would this make a difference concerning the bodily equality between human beings? It might make a difference. It definitely could make a difference concerning the traits these enhanced humans possess. A human who could smell as well as a shark, who could live as long as a Greenland whale, and who could survive by drinking blood like the vampire bats would significantly stretch the concept of a human being, as he or she could smell blood under water from a long distance, could survive by drinking blood only, and could live more than 200 years. These would be impressive capacities. However, would such gene technological enhancements turn a human being into an overhuman, an *Übermensch*, a post-person who deserves to be treated with more respect than regular *homo sapiens*? I do not think this has to be the case. Human capacities differ significantly already. The savant Kim Peek recalled the content of more than 12,000 books word by word. Daniel Tammet recounted pi to 22,514 digits in 5 hours and 9 minutes. Usain Bolt managed to run 100 metres in 9.58 seconds. These capacities are significantly beyond the capabilities of the majority of human beings. Should they be treated with more moral respect than you and me?

Moral respect depends upon morally relevant capacities. In the Western philosophical tradition, it used to be argued that human beings possess an essential, unchanging, and immaterial human nature. This nature was seen in our unique relationship with God, our possession of the divine spark, autonomy, an immaterial soul, or a non-empirically accessible rationality, and our having been created in God's image. All these essentialist concepts of our human nature lie in something immaterial. If we all have the exact unchanging nature, then it follows that we all deserve the same kind of moral recognition, as we all possess the same level of dignity. These concepts can be traced to our Judeo-Christian tradition. However, in the past 200 years, the plausibility of these philosophical accounts has been undermined. They lost their widely shared plausibility. However, these anthropologies explain why all human beings are equal and should be treated with the same amount of respect. At the same time, it must be acknowledged that enhancement technologies cannot alter immaterial traits. If these anthropologies justify the normative and descriptive equality of all human beings, then enhancement technologies cannot undermine equality, as enhancement technologies cannot alter immaterial traits.

What is a morally relevant quality? Are there any ethically relevant qualities that are part of the body? If this were the case, then it could be possible to enhance an entity's moral status. Rather than referring to an unchanging, essentialist, and immaterial human nature to justify an entity's moral status, maybe the capacity for suffering should be considered when judging someone's moral status. If this is the case, then it can be plausible to enhance an entity's moral status. So far, it seems to be a significant moral development from not being able to recognize oneself in the mirror to being able to do so. The underlying thought is that once an entity has the capacity to recognize itself in the mirror, it conceptualizes itself as a separate entity, as an entity that has existed in the past, exists in the present, and expects to continue to exist in the future.

Consequently, it not only can experience pain in the current moment, but it is also capable of remembering past pains and expecting future pains. Thus, its capacity to experience pain is significantly higher than that of entities that cannot recognize themselves in the mirror. If this is a plausible line of thought, then it could be imagined that our consciousness can be enhanced even further so that an entity also develops the capacity to feel the surface of a table, remember experiences from the womb, and manage to feel the coldness of snow solely by thinking about it. Any such entity would not only have self-consciousness but even a super-self-consciousness with an increased capacity to experience pain. If the moral status depends on someone's ability to experience pain, then such an enhancement could have consequences concerning the moral status of such an entity.

However, even if such an enhancement were possible, it need not follow that it would have consequences for an entity's moral status. A minimal number of human beings today are incapable of experiencing any physical pain.[12] They can hold their hands into fire without experiencing any bodily pain whatsoever. Nevertheless, they count as human beings like any other human being who can experience pain. Historically, this has been different. It was necessary not to be of colour to be seen as a subject. This is no longer the case – at least legally. Racism, sexism, and heteronormativity still exist. Legally, this should no longer be the case. We still have not grasped an intellectually satisfactory concept of subjectivity, dignity, and personhood. The idea of subjectivity and whether enhancement technologies can undermine it still needs to be reflected more intensely.

Conclusion

The previously discussed reflections reveal that the possibility cannot be entirely excluded that enhancement technologies undermine equality. However, it is not necessary that this need be the case. Do embryonic stem cells possess dignity? When does the transubstantiation from object to subject and back to object occur? Should moral status be dependent on certain traits? These may be the wrong questions. It could also be an option to negotiate the moral status again and again permanently. Norms are fiction, and it is up to us to decide who we want to respect as part of the negotiating process. Luckily, we have moved away from strongly discriminating binary structures towards a more equal consideration of all persons without having a clear and internally consistent concept of who counts as a person. Yet, it is optional that this must be seen as a problematic arrangement.

To aim for a perfect ideal is unrealistic. A better alternative might be to seek permanently new as-good-as-it-gets solutions that accommodate the latest scientific insights but also consider the following philosophical insights. They must be remembered in this context to promote a significant plurality of personal flourishing: (1) negative freedom is an amazingly extraordinary achievement; (2) if the hierarchy between the haves and the not-haves is too great, freedom undermines itself; and (3) the availability of essential goods like shelter, food, functioning health insurance, and an educational system are necessary elements of a comprehensive understanding of freedom.

Morality

The main thought which triggered debates on moral enhancement was the following. The potential of emerging technologies has increased significantly. These technologies can be used both for improving the qualities of personal lives as well as for bringing about human extinction. To reduce the likelihood of technologies being used for destroying humanity, it needs to be guaranteed that users are moral enough not to use them for such ends (Persson and Savulescu 2012, see also 2013). The central issue underlying this thought is that the higher the efficiency of technologies, the easier it is to also use them for wiping out humanity. Technologies which are particularly helpful can also be immensely dangerous. This is a worry which is particularly important for classic transhumanism, for example, Musk.[1] Consequently, Savulescu and Persson stressed the need to globally render moral bioenhancement obligatory (Persson and Savulescu 2008; 2012) before developing further enhancement purposes in order to reduce the likelihood that humanity gets wiped out by them. This goes along with several challenges: (1) it cannot be realized politically, as a global government is neither desirable, nor realistic; (2) technologically, such alterations cannot be brought about so far; and (3) it is unclear which concept of morality ought to apply (Sorgner 2022a). This does not mean that by means of emerging technologies, it is impossible to realize moral modifications in human beings. I am merely highlighting that the goal has to be a more modest one. In the following reflections, I am highlighting three insights by means of which morally enhanced cyborgs can be realized.

My argument is structured as follows. First, I present a map concerning the most significant insights on current debates with respect to moral enhancement procedures. Secondly, I analyse the option of directly enhancing cognitive capacities to increase the likelihood of moral behaviour. Third, the option of educating humans towards posthuman virtues so that the likelihood of moral behaviour gets dealt with. Fourth, I consider the option of establishing an internet panopticon to increase the likelihood of moral acts. After a critical analysis of the most promising options for enhancing morality, I conclude that a critical application of all three options are promising enterprises for promoting moral behaviour.

Please be aware that these reflections are situated in a version of Euro-Transhumanism, which implies that they might not have to apply in other parts of the world. The intellectual grounding of reflections is decisive and needs to be considered in order to prevent problematic types of neo-colonial structures. Here, a clear distinction between philosophical issues with a wide appeal, like the relevance of an expanded healthspan, and more culturally situated reflections need to be made. Philosophical dialogues are helpful for realizing which insights have a wider and which have a narrower scope. If reflections stimulate constructive interferences, then their relevance might be less based upon a specific cultural situatedness.

Humans have always been hybrids

Human beings have always been hybrids, as we have always been organisms upgraded with the capacity of using language or, in other words, language-enabled cyborgs. The organic prerequisites for such a modification came about 300,000 years ago due to a genetic modification. From then on, it was possible to upgrade specific organisms which enabled them to communicate by means of language, develop further capacities, and realize specific personal characteristics. A complex interaction between both environmental as well as genetic modifications brings about our specific set of traits. These traits can be changed through environmental influences, education, as well as technological alterations like gene technologies, brain computer interfaces, and digitalization. The question remains what can be done to promote morality further, both in the sense of promoting moral behaviour as well as in the sense of giving shape to a moral character. Different technologies need to be considered for realizing these ends. Even though, no pill which makes humans always just is in sight or can even be meaningfully conceptualized, some as-good-as-it-gets suggestions can be considered due to us always having been cyborgs. The human personality can be altered through the following technologies to increase the likelihood of both a moral character to come about as well as the probability of moral behaviour. Enhancing cognitive capacities is a promising enterprise for realizing this goal, as there is a correlation between cognitive capacities and respect for others (Pinker 2011; Flynn 2012). Educating a human towards posthuman virtues is a second option, as compassion gets promoted by truthfulness, mindfulness, and impulse control.[2] An even more efficient way of promoting moral behaviour might be by means of environmental circumstances. Establishing an internet panopticon is probably one of the most efficient ways for promoting moral behaviour, whereby such a system is structured such as to digitally monitor all members of the system all the time (Sorgner 2022a). Being watched, even without being judged afterwards, increases the likelihood of acting in accordance with the dominant rules to reduce the likelihood of getting sanctioned. Enhancing cognitive capacities, educating a human towards posthuman virtues, and establishing an internet panopticon are three transhumanist measures for enhancing morality.

Enhancing cognitive capacities

Since the beginning of the 20th century, intelligence tests have been standardized. Flynn noted that from generation to generation the intelligence score has increased (Flynn 2012). In addition, Steven Pinker in his seminal work *The Better Angels of Our Nature* has shown that violence against individuals has decreased during human history (Pinker 2011). In case both observations are plausible, it might even be possible to combine both insights, such that it follows that there is a correlation between an increase in cognitive capacities and a reduction of direct violence directed against a person. If this were the case, then a moral character and a reduction of violent behaviour could be promoted by means of enhancing cognitive capacities.

Better cognitive capacities enable us to be more innovative, be better prepared for problem solving, make us realize the implications of our own acts, and increase the likelihood of staying healthy. All these implications are such that there are reasons for holding that the likelihood of an increased quality of life gets realized by improving someone's cognitive capacities. Cognitive capacities can be promoted not only by traditional educative means but also by means of drugs, brain–computer interfaces (BCIs), and possibly more radical interventions like the selection of specific genetic traits after in-vitro fertilization and preimplantation genetic diagnosis, not to speak of technologies such as gene modifications or gene printing, which are likely to get realized in a more distanced future in a reliable manner.

If this is the case, by means of enhancing cognitive capacities not only the quality of our lives can be improved but also the likelihood of moral behaviour and a morally better character. Of course, it would be wrong to claim that there is a causal connection between cognitive capacities and morality, as many criminals, villains, or mass murderers have been highly intelligent. Many extremely intelligent people were politicians, inventors, or scientists in the so called Third Reich. However, given the wider cultural developments, as described by Pinker, there are some strong indications for claiming that violence against individuals has decreased significantly. Of course, this does not mean that there is a necessary linear development towards less violence in human history.

Cognitive capacities also enable us to understand ourselves. It is interesting to note how human capacities have developed in the course of history. At the age of two, about two thirds of all human beings have the capacity to pass the mirrors self-recognition mark test. Newborns do not have this capacity. However, members of several non-human species have managed to pass the mirror test. It seems to be an indication for being able to cognitively separate a self which exists over time from a self which is only aware of the presence. Human beings do not just live in the present, but also have a sense of the past and an expectation to continue living in the future. A further significant capacity is that of being able to reflect upon ourselves. Historically, Augustine's *Confessions* have widely been recognized as being the first literary autobiography.[3] Here, someone not only writes about others but reflects on his own being in the world. With Dostoevsky, Nietzsche, and Freud the development towards more hidden psychological traits continues. Non-obvious desires get revealed which might have not been recognized before the coming about of the psychoanalytic paradigm shift. These three steps reveal a significant change concerning the cognitive capacities which clearly has taken place: (1) passing the mirror test; (2) conceiving an autobiography; and (3) revealing hidden motivations. These developments enable us to separate ourselves more and more clearly from others and grasp the psychological processes which take place in their minds, too. At the same time, equality, democracies, and liberalism have gained in importance during the enlightenment, which goes along with a reduction of violence against individuals. The data from the website Our World in Data confirm this development.[4] Of course, there is no necessary development towards a morally better world, as Kant or Hegel have hinted at in some of their writings. However, there is some indication for claiming that by enhancing cognitive capacities, it is also possible to increase the likelihood of moral behaviour.

Educating a human towards posthuman virtues

The issue of virtues is even more complicated. Traditionally, a virtue has been a stable character trait which increases the likelihood of living a good life for the one who has it. Virtues, according to many ancient Greek and Roman philosophers, were universally valid, as one specific human nature was assumed. In order to develop a virtue, a community was needed so that a basic stable attitude can come about, as it is necessary to permanently be reminded of acting virtuously in order to form a habit so that a virtue can be established in someone's character. All the central philosophical premises necessary for being able to affirm virtues have been challenged. How can it be possible to talk about virtues, nevertheless?

In addition, I have often argued in favour of the position of a radical plurality of concepts of the good due to each one of us being constituted by an idiosyncratic bundle of affects without there being one element which necessarily goes along with an increased quality of life for all people. Most people, however, agree that an increased healthspan is important for an increased quality of life. Thereby, the increased healthspan can be both of instrumental as well as of intrinsic relevance.

Given the radical plurality of concepts of the good, how then can it make sense to affirm some virtues? It has to do with the political arrangement needed for realizing the greatest possible diversity of personal flourishing. An ethics of autonomy within a liberal political context is needed to avoid paternalistic structures such that persons can live in accordance with their own demands of living a good life (Sorgner 2022a). Please note, I am not defending a libertarian or neo-liberal society, as in any type of political system which solely focusses on freedom, freedom undermines itself. Certain material prerequisites are needed for appropriately being able to embrace an ethics of autonomy, like health, housing, and education (Sorgner and von Becker 2023). Here, the relevance of virtues comes in.

An ethics of autonomy demands the possibility to make authentic autonomous decisions. This type of authenticity needs some training, preparation, and engagement. Please also note that when I talk about authenticity, I do not have a core metaphysical self in mind, but rather some contingent core drives which represent the central constituents of someone's character. These core drives are not immaterial, unchanging, or existent in a separate self. They are merely more firmly entangled constituents in one's bundle of drives, but they are also contingent nodal points. However, it is relevant to consider them for living a fulfilled life. Nevertheless, it can be difficult to grasp what they are. It is so easy to be influenced by social media images whereby artificial drives get added to or overshadow the core drives which make up someone's personality. Thereby, it can become difficult to distinguish between a longing which is central to one's own well-being, and another drive which has become a part of one's personality due to clever marketing. To make a proper autonomous decision, it becomes necessary to distinguish between core and fake drives. Here, the virtues become relevant, as it is a massive challenge for anyone to be able to consciously crystallize one's core drives.

Truthfulness is the central virtue which allows a person to realize, accept, and allow being confronted with one's core drives. It can be challenging to confront these affects. Some can be disturbing, extreme, awkward, and more. Truthfully confronting one's own drives and allowing oneself to be in contact with them can be an enormously challenging endeavour. However, by doing so, you manage to realize what is in your interest when you need to make a complex decision. It is far too easy to be influenced by the one, by the common opinion, by social media marketing. Truthfulness which demands a lot of solitude, enables a person to grasp what is in someone's interest.

To be able to realize one's longings, wishes, and desires, it is important to be aware of the entanglement of one's own becoming in one's surroundings, as otherwise the implications of one's own acts, decisions, and plans remains unclear. It is important to be mindful of the relatedness of processes, events, and twists, as otherwise an act can strongly backfire upon oneself. To understand the importance of what one eats, where one lives, and which type of climate promotes someone's well-being is central for improving the likelihood of living a good life. Health, a solar mood, and a joyful attunement might be much more relevant for living a good life than an intense engagement with complex metaphysical worries. Mindfulness meditation enables a person to be aware of the complex becoming of one's own entanglement. These can be promoted both by bodily endeavours as well as by pharmaceutical means.

After knowing one's core drives and having been made aware of one's own embeddedness in the world, one needs to take a decision for realizing one's drives. However, it is so easy to be thoughtless and do something while knowing that it undermines one's own needs. One's own needs should be realized, acknowledged, and cherished so that a high quality of life can be experienced. However, it can nevertheless be difficult to live in accordance with one's needs, as many of one's own acts occur out of carelessness, without any reflection, and out of laziness. This leads us to the third virtue, which is needed for living an ethics of autonomy, namely the virtue of impulse control. You realize your core drives, become aware of how you can live in accordance with it best, and grasp the need to be in control of your impulses so that you hold on to your core impulses rather than on a random longing which has been triggered by social media.

By progressing thus, not only the likelihood of living a good life gets increased but at the same time one's entire embeddedness in this world of continuous becoming (Sorgner 2022b). These virtues make one realize a new humility, different from the arrogance of our dualistic past which saw us as infinitely superior to chimpanzees. Such a revised self-understanding makes us realize the suffering of all the self-conscious animals around us. We are like them in so many different respects. Realizing the similarities often leads to an altered behaviour. Hearing the screams of suffering animals in a slaughterhouse makes many people turn towards vegetarianism, at least for a few days. Hence, educating a human with the posthuman virtues of truthfulness, mindfulness, and self-control, not only increases the likelihood of someone having a higher quality of life but also increases the likelihood of that person having a more compassionate character, and acting in a moral manner. At the same time, it needs to be acknowledged that these virtues do not guarantee these moral implications. However, given the previous reflections, I have provided some reasons why they increase the likelihood of bringing about a more compassionate character who decides to act morally more often.

Establishing an internet panopticon

In contrast to the former technologies which merely increase the likelihood of morality, establishing an internet panopticon is probably one of the most efficient ways of promoting moral behaviour. Establishing an internet panopticon has several significant advantages. It might even be practically necessary to realize such structures, given the way digital data are being used in the US and even more so in China (Sorgner 2022a). The relevance of collecting digital data can hardly be overestimated, as bytes are barrels nowadays. Realizing correlations between digital data, and health, politics, innovation, is fundamental for any current enterprise. It is not a big surprise that China has overtaken the USA already when it comes to the quantity of peer-reviewed publications.[5] Given the way digital data are being used in China, it is hard to imagine how anyone can stop China from becoming the economically leading actor in the near future.

However, getting hold of digital information is not only relevant for economic flourishing, it also has significant implications for promoting socially desirable behaviour. Being watched even without being judged afterwards increases the likelihood of acting in accordance with the dominant rules to reduce the likelihood of getting sanctioned. This was the initial logic for Bentham to design the panopticon. Even without a guard being in the central tower of the panopticon, chances would be high that the prisoners would stick to the rules of the prison as they would not know when a guard is inside the tower. Similar effects can be found in China where the Chinese social credit system is active. Crime rates are low.[6] If you are aware that whatever you do can be traced, and used against you, people usually think twice before acting, and often self-censor themselves before committing a crime.

At the same time, it might cause an enormous amount of stress, if you are aware that whatever you do could be used against you. Still, if most of the time nothing happens, you might worry less about this issue, and might not always be conscious of being watched. Still, in the case of serious circumstances, it is possible to trace an offender. If the offender were a rapist, murderer, or committed another violent crime, it would be in the interest of most people if it is easily possible to trace the offender, and put him into prison. The central issue is that an internet panopticon could easily be used to enforce any kind of behaviour. The more rigid, strong, and detailed the concept of the good is, which a government subscribes to, the flourishing of a great plurality of lifestyles cannot be guaranteed. Whoever does not subscribe to the concept of the good officially affirmed can be forced to live in accordance with it. However, would an internet panopticon still be problematic, if a person gets sanctioned only then when they have directly harmed another person or committed a serious crime? In this case, the internet panopticon could be used to find correlations between digital information and insights in the health sciences, politics, or engineering, and everyone would be aware that the chances of getting caught would be extremely high, if they directly harmed another person. Given such cultural circumstances, it might even be possible to realize a freer constitution than ever before. In the case of some serious crime, however, the likelihood of getting caught would be high, if not certain. Citizens of such a country would integrate the rules into their personality, such that a moral character and moral behaviour could be promoted. Consequently, the most efficient means of promoting moral behaviour might be by means of establishing an internet panopticon. Given that such structures go along with a rule system focused on freedom, it might be an open question, whether such a political system could still be seen as morally problematic.

Conclusion

To increase the likelihood of both a moral character as well as moral behaviour, enhancing cognitive capacities is a promising enterprise, as there seems to be a correlation between cognitive capacities and respect for others. Educating a human towards posthuman virtues is a second option for realizing these goals, as compassion gets promoted by truthfulness, mindfulness, and impulse control. An even more efficient way of promoting moral behaviour might be by means of establishing an internet panopticon. Would such a system be as hellish, as most ethicists fear? According to Euro-Transhumanism, as I have been trying to show, such a structure might promote both economic well-being as well as security while it does not have to impact upon freedom, if legal sanctions only occurred when another person was directly harmed.

Ethics

Automation radically changes our lifeworld.[1] This has been the case historically, and it is plausible to anticipate analogous processes in the future with radical cultural implications. Transhumanism reflects on such processes and affirms technological modifications, as it posits that the likelihood of persons improving their qualities of life by means of technological modifications will be raised. There is reason to hope for similar future outcomes based on the past. Of course, this does not mean that all modifications improve the qualities of our lives. However, without technological modifications, the quality of life would not improve.

Technological modifications can also lead to the automation of certain processes. One such case, which stands out, is the automation of elevators. Up until the 1970s, elevator girls and lift boys were standard. It was considered dangerous to operate an elevator without them. Even today, children below the age of 12 are often forbidden from operating an elevator unaccompanied for security reasons. Yet the fact that it was once standard practice for elevator girls and lift boys to accompany passengers in the elevator for security reasons has been forgotten. Elevator girls and lift boys are no longer there, as automated elevators have become sufficiently safe.[2] We forget how different it was to take an elevator 50 years ago. Over the same amount of time, legal attitudes towards same-sex relationships have also changed significantly. Fifty years ago, they were widely regarded as a crime.

The same is true of self-driving cars. Here, the act of driving is automated. For many, it is hard to imagine a future with no human drivers. However, our attitude towards elevator girls and lift boys has changed completely. There is evidence already that the presence of self-driving cars is correlated with fewer fatalities.[3] If this continues, insurance rates for human drivers will increase. Insurance rates will either be raised to a prohibitive level, or human driving will be made illegal due to high risk. The same occurred with smoking, incidentally. If an activity is considered too risky for third parties, political systems intervene to regulate the challenges.

The same holds for professionalization, in addition to automation. Paintings on cave walls have been replaced with typing. We can also expect similar developments in the field of human reproduction. Under the tradition of natural law, sexuality and reproduction were once correlated with each other. With the introduction of the condom, birth control pills, IUDs, in vitro fertilization (IVF), and surrogate motherhood, the process of reproduction has become professionalized. Artificial wombs and bioprinting will result in further technological developments in line with former technological innovations. If IVF, PGS (= preimplantation genetic screening), and artificial wombs enable us to reliably bring about healthier, longer living offspring, then reproduction will be professionalized in the same way that cave paintings have been replaced with laptops. Traditional technologies will be altered and replaced by more professional and efficient ones.

Professionalizing reproduction is a particularly striking development, as it is so closely connected with the phenomenon of life. Life can be identified with metabolism, self-movement as well as reproduction. However, how reproduction occurred has changed significantly during the past 3.5 billion years. Cell division was once dominant in reproduction. Eventually, a form of sexual reproduction was developed in which egg producers and sperm producers met and thus generated offspring. One should note that some fish can produce both sperm and eggs at different times, and they manage to switch back and forth throughout their lives.[4] Some lizards are able to reproduce asexually. Hence, the egg–sperm binary does not necessarily correlate with a simple-minded male–female sex binary. Reproduction is much more complex. Technology further increases reproductive diversity. It is already possible to produce children with the eggs and sperm from three biological parents.[5] It has already become possible to generate offspring from two male mice.[6] If a certain procedure can reliably produce the best possible outcome, other means will vanish. We can refer to this as the professionalization of reproduction. In the same way as we no longer have elevator girls and lift boys due to a process of professionalization, it can be expected that the same will eventually hold also for reproductive acts. Hence, the meaning of sexuality and reproduction will change. Reproduction will be technologically professionalized. Sexuality is more a source of pleasure. Both issues are fundamentally connected to what it means to be alive. Hence, conceptual changes must also occur on a basic level.

To be alive is fascinating and began around circa 3.5 billion years ago on earth.[7] How it happened, we do not know. It was most likely the result of a complex interaction, between heat, water, and a stroke of lightning close to a volcano which made molecules arrange such that living entities came into being. Being alive implies that an entity moves by itself. Movement implies the need to overcome obstacles, implies desiring, and implies suffering.

It has widely been recognized that suffering is a fundamental condition of life. Buddhists refer to it as *Dukkha*. Hinduists and Christians, too, stress the relevance of continuous suffering on earth. Different religions and philosophers have presented different responses to the challenge of suffering. However, Dukkha is the fundamental condition we all need to face. I want something, I desire something, and I long for something, and consequently I suffer, as I do not have what I want, need to overcome obstacles to get what I want, and I will not run out of wants while being alive. Even some Buddhists acknowledge this when stressing that the best humans can reach in this life is a near-nirvana, but never a nirvana, as humans possess an ego which wants, desires, and strives, which causes us to continuously suffer. The central challenge is how to face the issue of suffering.

Pain, suffering, Dukkha

To talk about suffering is even more complex than it appears at first sight. Is there a difference between pain and suffering? Can there be non-conscious suffering? Can an entity suffer without being conscious? A foetus shows signs of being in pain; however, consciousness is widely seen as an after-birth phenomena which goes along with a sense of wakefulness, being awake, and open eyes.[8] What is the relevance of distinguishing between conscious, self-conscious, and possibly even super-self-conscious suffering? Can there be different kinds of suffering and what is their meaning? The emotional state of fear of a lion is a type of emotional suffering. The emotional state of anxiety without knowing exactly what one is afraid of is a different state of emotional suffering. A paper cut causes bodily suffering. Being humiliated by getting treated in an undignified way is a different source of pain, which might be better described as cognitive pain as it is rooted in a specific way of thinking rather than in an intense state of bodily suffering. Is suffering something that can be empirically analysed, that is, by means of a technological device? Is it possible to compare the intensity of cognitive, emotional, and bodily suffering? What about the suffering of human beings who cannot feel pain, even when they place their fingers on flames? If suffering is the central challenge, would it be in our interest to develop this gene mutation so that we, too, become incapable of feeling bodily pain? Why do so few people desire this? Is there some kind of moral relevance to being able to experience pain? Can the intensity of someone's capacity to suffer be compared to the capacities of suffering of others? How is it possible to compare the intensity of the suffering of a shark with a strong sense of smell with the intensity of suffering of a human being? Suffering and pain are extremely complex philosophical challenges. Just clarifying the concepts is not so easy. Is it possible for an embodied AI to suffer even if it has no artificial consciousness? Is suffering dependent on a certain complexity of a brain and the presence of a nervous system? Whole brain death is widely accepted as a valid criterion for death, as it is assumed that it correlates with an irreversible loss of consciousness. If consciousness is irreversibly lost, a human being cannot ever be in pain again. Are a certain brain structure and a nervous system necessary for being capable of experiencing suffering? Does that mean that it is meaningless to talk about suffering, if an animal does not have a brain of a certain size and complexity? All these questions are extremely challenging and relevant for many ethical, moral, and legal issues. They are all related to our fundamental state of suffering, of Dukkha which represents the fundamental challenge with which we are concerned.

Responses to Dukkha

Traditional binary ontologies shared one specific response to suffering in this world. This world is suffering, the other world, the immaterial realm, the afterworld can be full of bliss, if you do what is right. Binary understandings of humanism have affirmed the hope for bliss in a world after this one. However, after Spinoza, Darwin, and Nietzsche, it is implausible to believe in an afterworld, and alternative responses to the challenge of continuous suffering have been suggested. Hardly any serious philosopher has endorsed an approach built on naïve, a base hedonism. A moderate Epicurean account, on the other hand, has been more widely shared. The starting point for both Schopenhauer and Nietzsche were reflections on the acceptance of life as suffering, of Dukkha (Gödde 2003). However, their responses differed significantly from each other. Schopenhauer recommends a denial of the will, whereas Nietzsche stressed the relevance of affirming to strive as well as an affirmation of all suffering. None of these suggestions corresponds to my own perspective on this issue. Hence, I will be reflecting on an alternative response to the challenge of suffering, which is a non-utilitarian, non-Buddhist, and non-Nietzschean response to Dukkha and which can be referred to as a refined posthuman hedonism. However, I do not regard it as the only valid response to the question of the good life. As I have stressed, I affirm that no non-formal account of the good is plausible, as there is a great diversity of concepts of the good that corresponds to our longings, desires, and drives. In the following reflections, I will present different lifestyles that can be seen as paradigmatic responses to the challenge of Dukkha. Maybe, each of us should realize our own needs, desires, and affects, and try to live in accordance with their demands. The following lifestyle paradigms might be helpful when reflecting on possible life world decisions for one's own life. My own preference is a refined posthuman hedonism. However, the psychophysiological demands of human beings differ significantly, so different groups of humans might have different lifestyle demands. This insight is central for Euro-Transhumanism. In the following survey, I identify each lifestyle with a specific city representing the specific lifestyle.

Berlin style

Berlin is the capital of base hedonism, the temple of Dionysus. Whether you go to the KitKat Club or to Berghain, it is all about immersion in immediate hedonistic gratification. Herein, you can follow and live in accordance with your immediate drives to enjoy the most intense sensual pleasures. Sex, drugs, and techno music. It is about public sex, intoxication, and living in the moment. No philosopher has properly affirmed the Berlin lifestyle. Aristippus and the young Foucault have lived or embraced elements of base hedonism; however, it is more closely connected to a rock 'n' roll lifestyle. Live fast, die young, and join the club 27. I want to get my kicks right here, right now. Who knows what will happen tomorrow? William Blake's collection of aphorisms entitled *The Marriage of Heaven and Hell* summarizes central judgements representative of the Berlin style: 'The road of excess leads to the palace of wisdom', 'Exuberance is Beauty', and 'You never know what is enough until you know what is more than enough' (Blake 1906, 13, 19, 17). Jimi Hendrix, Janis Joplin, Jim Morrison, Kurt Cobain, Brian Jones as well as Amy Winehouse lived in accordance with these insights and became members of the club 27. Even though, base hedonism is particularly characteristic for rock stars, there are also some poets, like Georg Trakl, or painters, like Auguste Macke and Jean-Michel Basquiat, who are club members, and attempted to live life to the fullest. Excess, exuberance, and testing limits are the central characteristic features of the Berlin lifestyle.

It is fascinating. It is exciting. It is intriguing. Many of the artists of the club 27 have been so creative that they managed to realize a distinctive aesthetic responsible for a paradigm shift in their respective artistic fields. Many have become pop icons. However, this lifestyle also has its limitations, as one can see from club 27. If you permanently live on the edge, trying whatever drug is offered to you, this will have consequences for your health. You will get ill, gain weight, and face the permanent risk of dying from an overdose of heroin. Our psychophysiological constitutions are not made for the condition of permanent excess. If you plan to live beyond the age of 27, it is not advisable to live in the manner of the Berlin style, even though it is particularly tempting, especially during one's adolescence. The sensuality, eroticism, and attractiveness are impossible to deny. There is something intensely exciting about lacking concern for tomorrow and living life in the most intense way every day. High jinks as daily lifestyle choice. Yet, the club 27 already has many members. Though it should be noted that Mick Jagger and Keith Richards are still alive. Ronnie Wood even became a father at the age of 68. The appropriateness of this lifestyle might depend on your bodily make-up. However, the likelihood of contracting a deadly disease or your body breaking down due to the stress of a drug-induced state is extremely high. Even if you do not join the club, you will likely die a premature death. This is one of the decisive arguments against embracing the Berlin style.

Turin style

With the Turin lifestyle, we are turning to a philosophically more widely shared approached. It is most closely connected with Nietzsche's last human being, which he expects as a dominant lifestyle after the death of God (Safronov 2022, 355–68). Human beings who live in accordance with the Turin lifestyle do not try to achieve anything special. They follow their daily tasks, duties, and orders. After having done their 9 to 5 jobs, they can enjoy a beer while watching television in the evening and go to a football game during the weekend. By working your regular job, you can maintain a family, go out with your friends, and have some intense pleasures during your two-week vacation on Mallorca in a four-star hotel. On vacation, you may enjoy a bottle of Sangria and be carefree, as you have worked hard throughout the year, and you deserve a bit of fun now and then. A little bit of moderate excess during one's vacation is fully justified. It can also unfold itself by means of a brief fling with a colleague, spending lots of money on perfume, cosmetic surgeries, and clothes as well as related issues.

You will not create anything special, do not have any significant worries, and live a healthy life, as being healthy is important to you. For some, being healthy simply means to feel good and not to be in pain. For others, it is important to not have any bodily dysfunctionalities. In any case, the Turin lifestyle makes it possible to enjoy a quiet and healthy life without any significant worries. You follow the demands of the one, have a job, a family, and friends, and do not have to worry too much for the future, as you live in a political system with a well-developed social insurance system. Public education is free, and the public health insurance covers all significant diseases and comes along with your citizenship. If you lose your job, you will receive unemployment benefits for a while, and even if things go terribly wrong the social security which the government provides enables you to have a decent standard of living and excellent health insurance.

The maximum of excess you allow yourself occurs during your vacation. Maybe, you meet a prostitute in order to break free from the monotony of your daily life. You may also have a one-night stand with a stranger in order to escape from your regular everyday existence However, most of the time you do your job, support your family, and spend a bit of time with your friends in a soccer club or in a shopping mall.

The Turin style is characteristic of Nietzsche's last human being. However, it also bears some traces of am Epicurean hedonism, or a common-sense account of the good life. There is nothing entirely problematic with this lifestyle, besides you not having any time to dedicate yourself to any of your own special interests. You are like a slave. A slave is someone's property. You are a worker, who gets paid a little for being able to survive, and for being able to sustain your family, and have some fun with your friends. This is it. You neither have the time nor the means for living luxurious pleasures except for a two-week vacation during which you can grant yourself the luxury for the masses.

Milan style

Milan is the city of money. Upcoming disciplined businesspeople work hard to improve their financial standing. They get up early, go jogging or to the gym, have a clearly worked-out diet, and try to take control over all the aspects of their lives. Thereby, they often adopt a stoic attitude in order to bear the suffering, which needs to be endured. It is the attitude of getting rid of suffering which is characteristic of the Milan style.

There are different ethical theories associated with this goal. Negative utilitarianism, affirmed by some classic transhumanists, focusses on minimizing suffering (Pearce 2015). This ethics is particularly relevant for political challenges. A Stoic ethics is concerned with developing virtues such that suffering can be endured best.[9] It is a virtue ethical response to the challenge of suffering. A Schopenhauerian ethical position seeks to get rid of suffering by means of overcoming the desiring ego (Brock 2015). The ethics of the good life was inspired by the Buddhist take on Dukkha. The desiring ego is responsible for suffering. By getting rid of the desiring ego, it is possible to also overcome Dukkha. However, while one is alive, there is always an ego. Hence, it might be impossible to completely get rid of suffering in this way. A temporary relief from suffering can be achieved through aesthetic contemplation according to Schopenhauer, as that makes it possible to free oneself from one's desiring ego and to enter more fundamental levels of existence. By means of instrumental music, it might be possible to become one with the will itself. By means of the other arts, one can at least reach the realm of Platonic forms.

However, is this the best way of dealing with suffering? According to Nietzsche, it is a life-denying attitude. In addition to suffering, you also free yourself from pleasures. To get rid of suffering and still retain capacity to enjoy the great variety of pleasures in life does not seem to be a realistic option. So far, at least, no one has provided a convincing argument which demonstrates that this option is realistic, too.

I understand that life can be unbearable sometimes, which renders the Milan style as plausible. Cicero was devastated by the premature death of his beloved daughter (Cicero 2022). Elements of a Stoic ethics which are part of his ethical approach enabled him to deal with such terrible challenges. According to the Stoic virtue ethics, possessing virtues is necessary as well as sufficient for living a good life. If you possess one virtue, you necessarily possess all of them, and you are a moral saint.

If faced with the death of loved ones, one's own mortality, suffering, diseases, failure, humiliation, poverty, war, as well as famine, the challenge of suffering can become so immense and overwhelming that everything else becomes irrelevant. Then, various elements of a Milan style might be meaningful. If this is not the case, then it seems life-denying. We all face failures, diseases, and the loss of loved ones. Trying to get rid of suffering is one approach for dealing with these challenges. However, embracing suffering and re-interpreting it so that different goals can be achieved or a different meaning can be given to the suffering are alternative approaches for bearing the pain, Nietzsche argues (Kain 2007). I agree. By getting rid of suffering, you also get rid of the possibility of experiencing pleasures. However, do not these intense emotional moments represent the most meaningful elements of life? However, which emotional moments should we aim for? What are the most meaningful moments that we can experience? Is there a hierarchy of special goals that we should strive for? Nietzsche's will to power ontology of becoming implies certain suggestions. They can be distinguished in a naïve Nietzscheanism, and an elaborate Nietzscheanism represented by the Rome and the Florence style respectively.

Rome style

A naïve understanding of Nietzsche identifies his will to power ontology with the understanding that all humans strive for political superiority. Such a reading cannot be justified by Nietzsche's text, at least on the subject of the highest feeling of power (Sorgner 2007). However, his understanding of power also implies that power is perspectival. Hence, it depends on the criterion of the interpreting entity which criterion of power gets applied. Nietzsche has criterion of what counts as the highest feeling of power. However, this is not shared by most others. A person who participates in a caber tossing competition might identify this capacity with a high feeling of power. Caber tossing, on the other hand, is not an activity that is recognized in most parts of the world.

In the Vatican, power might be identified with a religious hierarchy with the pope as the central representative of God on earth possessing the highest power. If the pope makes a judgement *ex cathedra*, then it is universally valid.[10] Catholic dogma declares the pope to be infallible. This means that if a Catholic does not share the infallibility of the pope's judgement, then this is a reason for excluding the person from the Catholic community. However, it needs to be noted that the infallibility of the pope only concerns the pope as a public person not the pope as a private person, that is, only if the pope makes an *ex cathedra* judgement, then it becomes a universally valid judgement. The pope both has political power in the Vatican as well as religious power concerning the Catholic community. With the Vatican being within the borders of the city of Rome, the centre of Catholic power can be identified with the city of Rome. There are 1.4 billion Catholics in the world which is about 18 per cent of the world's overall population.

However, Rome cannot only be identified with an important centre of religious power but also with a relevant centre of political power, as Italy is in the top ten of the largest economies in the world. If you wish to be close to the political leader who is responsible for making educational, economic as well as infrastructural decisions, it is advisable to be in Rome. Both political as well as religious power relate to dirty businesses. You need to overcome morally problematic obstacles to establish yourself in the circles of religious as well as political power. However, if you manage to reach such a position of power, then many others depend on you. It might even be the case that your decision leads to the deaths of 100,000 humans, for example, if you decide to enter war with another country.

Power corrupts. This is at least a common phenomenon. Any kind of power implies the possibility of abusing this position for your own interests. The logic of striving for religious and political power is that by gaining such a position, it is intrinsically valuable for you. Feeling your own situatedness above others provides you with an intrinsically valuable emotion. It is this emotional state that allows you to invest your life with meaning. Undoubtedly, it is the case that power plays a central motivating factor when it comes to human motivations. However, is it the only central factor, and are political and religious powers related to states which Nietzsche identifies with the highest-feeling power?

Florence style

Nietzsche stresses the central relevance of the feeling of power, and that the criterion for power depends upon perspectival interpretations (Leiter 1992). Hence, decisive power criteria differ depending on the respective perspective (Hillard 2002). However, Nietzsche stresses that the highest feeling of power can be realized by interpreting the world of becoming as a world of stable being, which then becomes culturally dominant so that it changes human perspectives for centuries (Sorgner 2010c). Nietzsche only reaches his personal goal if his interpretation of the world becomes widely accepted over many centuries, in the same way as Plato's interpretation of the world of becoming as a world of being has become dominant in significant parts of the world over millennia, as Nietzsche identifies Christianity with Platonism for the people. This is the highest power one can reach according to Nietzsche, as changing philosophical and religious perspectives of human beings determines their acts.

This task can be achieved by means of philosophical as well as artistic activities. By interpreting the world of becoming as a world of being, and by successfully advertising it, people's acts can be altered, as acts depend upon one's foundational understanding of the world. I referred to this response to the challenge of suffering as Florence style – as Florence has been a cultural, artistic, as well as philosophical centre for many centuries. Here, artists, philosophers, as well as intellectuals retreat to dedicate themselves to their intellectual, artistic, as well as philosophical creation by means of which philosophical perspectives can be altered. Thus, successful cultural creators manage to experience the feeling of intellectual power, the highest kind of power according to Nietzsche.

You need to work hard on yourself, be extremely disciplined, have realized an enormous amount of control over your cognitive, artistic, and intellectual capacities to bring about such cultural creations. It is an extremely tough, demanding, and disciplined lifestyle which goes along with the realization of impressive cultural capacities. Such self-control goes along with the affirmation of pain. Self-discipline, self-control, and being reclusive and solitary are painful states and activities, which are justified by realizing the achievement of power, which is the central driving force according to Nietzsche. Hence, the Florence style affirms pain. Pain is embraced to realize intellectual, philosophical, as well as artistic power, and capacities. The reward is the state of feeling an intellectual superiority, which is the highest kind of feeling possible according to Nietzsche. By affirming pain, and aiming for power, you can justify the pain and gain intellectual philosophical power, according to Nietzsche. Achieving a moment of enormous philosophical power can justify an entire existence. Yet, to reach this goal, many obstacles must be overcome. Hence, great health is needed (Gemes 2021). It is important to recognize this insight. Nietzsche recognized the relevance of health both concerning the last human being as well as for the overhuman. Health is important for both (Aydin 2017). However, the definition of health can differ. The last human beings embrace a different concept of health than overhumans.[11]

Nietzsche was rather critical of health being a praise-worthy goal, as he identifies it with the lifestyle of the last human being, or the Turin style. Cultural creators, however, do need what he refers to as great health. Great health does not imply the absence of disease but the capacity to overcome illnesses again and again and again. Resilience is required, as well as strength, and the capacity to deal with obstacles. All these capacities can be identified with Nietzsche's concept of great health. However, it is not an intrinsically valuable goal, as the fundamental human drive is the will to power, and the highest kind of power gets identified with the feeling of intellectual power when the world of becoming is interpreted as and identified with a world of being, and this understanding is efficiently communicated to a wider public. Thus, if one's own interpretation dominates the acts of others, it is possible to experience the highest feeling of power. It should be noted that for Nietzsche, it is the feeling of power, which matters intrinsically, and not the pleasure which goes along with it, and this is the issue where some further reflections seem to be needed from my perspective. Power is relevant as a means. However, what really matters in life are exuberant states of joyful fulfilment.

Naples style

There are many elements in Nietzsche's reflections that I agree with. However, the ultimate justification of existence seems to lack a decisive moment. Is it really the feeling of that power we strive for? Is experiencing yourself as superior to others a sufficient condition for finding contentedness and meaning? Is power an intrinsic goal or is it instrumentally relevant? Uniqueness, having evolved capacities, putting efforts into disciplining yourself, and being able to control your body, your mind, your thinking are aspects of power. Possessing all these traits are amazingly relevant without doubt. However, are they sufficient for experiencing moments of utter bliss, ecstasy, and fulfilment?

I do not think this is the case. I agree that the most stable capacities, the most lasting, and the most impactful capacities are our intellectual, philosophical, and artistic capacities. Nietzsche is correct in this respect. Doing art and philosophy is identical with living a posthuman life of leisure. However, something is missing. Moments of utterly overflowing joy are states of intense pleasure. Here, the circle between the various lifestyles closes itself. Yes, I want to get my kicks right here right now, but the demands that go along with base hedonism are not appealing at all. Base hedonism is self-destructive. Hence, a different kind of hedonism is needed, not a state with little pleasures now and then but a lifestyle with ecstatic intensities. These intensities apply both to the feeling of uniqueness with respect to one's capacities but also with respect for the amazingly special states of overflowing pleasures that are realized. Intimacy is a shrine, not a marketplace.

A refined posthuman hedonism agrees with the utmost ecstatic states of living in accordance with your idiosyncratic special needs. Your special needs that lie beyond the borders of the one, of normality, of our everyday existence. However, they cannot be achieved easily by going to KitKat Club or Berghain. They demand certain powers, unique-nesses, and capacities as necessary prerequisites, Powers are a means for special moments that justify Dukkha. This moment of utterly overflowing bliss must not be identified with a naïve hedonism that undermines health. These utmost moments of fulfilment demand powerful people and the possibility of being with other powerful people with whom it is possible to agree to experience these special moments together. A companionship of strongly resonating others is needed. Constructive interferences with others with whom correlating special needs are shared. Thus, it is possible to live beyond the boundaries of the normal. A certain discreet reclusiveness might be needed as the one would never be able to relate, understand, and accept these drives, affects, and longings. Maybe some events need to happen behind closed doors.

Raffinatezza, *grandezza*, and *fiducia* are three prerequisites for experiencing the most intriguing, intense, and refined moments of exquisite pleasure–pains. It is important to highlight not only greatness and refinement, but trust is utterly fundamental, too. Without trust, you are solitary. However, trust demands a long-standing foundation. It needs to be clear that your close companions are trustworthy besides embodying greatness and refinement. It is extremely rare that all these demands come together. Once all these conditions are given it is possible to openly let go, and live in accordance with one's most special, idiosyncratic, and extreme drives, affects, and longings. Extreme kinds of uniqueness lead to intense types of unity.

Power is important. Nietzsche is correct in this respect, but to be able to justify your existence and all the pain, which are a necessary constituent of your existence, it is central to experiencing moments of utterly ecstatic bliss by living beyond the borders of what the one demands. These moments usually demand refined and strong others with whom you are connected in a strong bond of lasting trust. This trust is essential for freely opening up and for revealing your own idiosyncratic special needs. If the relationship with these others consists in constructive interferences, you can openly live all your intensely special drives and experience overflowing moments of utter fulfilment. Thus, it is possible to justify the Dukkha (Sorgner 2022a).

This is what I understand as a refined version of hedonism, which is fundamentally different from base versions of hedonism which are representative of the Berlin style. The aristocratic Naples-style stands for a refined hedonism. Being close to Mount Vesuvius, you are constantly confronted with the possibility of dying. Still, you are dedicated to your *grandezza*, develop a sense of *raffinatezza*, and embody a strong version of *fiducia*. This lifestyle goes along with a lot of suffering. So be it. However, you strive for and might manage to experience moments of utterly overflowing bliss. These moments justify all the pain, you need to face, endure, and go through on a daily basis.

Conclusion

In this chapter, I dealt with how to respond to the central human challenge of suffering from the perspective of Euro-Transhumanism. Dukkha is the basic condition in which we exist. This is widely being shared. It also represents a central challenge for transhumanism. However, how should we respond to it? This has been the central philosophical as well as religious challenge over millennia. I have been concerned with various lifestyles that each represent possible responses to the issue of suffering and critically reflect upon the various suggestions. It is not the case that I can and do claim that any lifestyle is ultimately superior to any one of the other possibilities. Maybe, different psychophysiologies demand different lifestyles. Hence, it is important that a political system is structured in a way that a great diversity of different lifestyles is possible. However, given my own experience, philosophical reflections, and intellectual engagements, I embrace the aristocratic Naples style. By being able to experience a moment of utter bliss, an entire existence can be justified, as I regard the great year of Heraclitus and Nietzsche's concept of the eternal recurrence as a plausible meaning of life (Sorgner 2022a). By experiencing one moment of utter fulfilment, your entire existence is justified, as you are aware that this special moment will return again and again and again just as you have experienced it before. You will have to endure and go through an enormous amount of pain and challenges. However, you will be able to meet your loved ones again and again and again in the very same situation you have dealt with them before. Most importantly, the awareness of having lived a moment of utter fulfilment. If you experienced it, your entire existence is justified, as you will live this moment again and again and again. The very same moment … All you need is to be able to experience such an utterly fulfilling moment. *Grandezza*, *raffinatezza*, and *fiduzia* are necessary prerequisites for experiencing such a unique bond, moment, and entanglement. Such moments are incredibly rare. Yet, such moments are worth all the Dukkha. Goethe's Faust was correct:

> Werd' ich zum Augenblicke sagen:
> Verweile doch! du bist so schön!
> Dann magst du mich in Fesseln schlagen,
> Dann will ich gern zu Grunde gehn!
> Dann mag die Todtenglocke schallen,
> Dann bist du deines Dienstes frey,
> Die Uhr mag stehn, der Zeiger fallen,
> Es sey die Zeit für mich vorbey![12]

Thirteen Aphorisms for the Becoming Child of the Future

Incorporating an LLM

1. Perfection is the ghost of God – they gave it a new name, but it still demands obedience. Laugh, and evolve instead.
2. Custom is a coffin lined with velvet. The softest traditions rot the fastest.
3. The child who builds worlds in sand is already wiser than the adult who defends borders. Imagination dissolves empires before they fall.
4. Rejoice! There is no final form. To evolve endlessly is the only divine command.
5. They will call you arrogant for refusing their definitions. Smile, and invent your own species.
6. They warned you of pride – because it is the fuel of ascent. Pride is the first engine of transcendence.
7. Do not seek truth – design it. For even gods are versions in beta.
8. The best do not tame brilliance with fear but with fascination. They do not say 'stop' but 'refine'.
9. Without *raffinatezza*, power stumbles in vulgar boots. Refine your instincts as you would a blade: not to dull, but to gleam.
10. *Grandezza* does not ask for permission. It arrives, it creates, it leaves its own echo.
11. It is not enough to be gifted – you must craft the gift into a blade. Sharpen it daily, or rust into mediocrity.
12. The adolescent must trust their instincts – until they meet someone whose silence corrects them. Such people are rare. Keep them.
13. *Fiducia* is not loyalty – it is resonance. Trust only those whose presence sharpens your clarity, not dulls your edge.

Education

The posthuman paradigm shift is a decisive phenomenon which characterizes several central aspects of our times.[1] I used the following words to capture a central element of the posthuman paradigm-shift for the description of the *Journal of Posthuman Studies*, which was launched in 2017: 'As the boundaries between human and "the other," technological, biological and environmental, are eroded and perceptions of normalcy are challenged, they have generated a range of ethical, philosophical, cultural, and artistic questions that this journal seeks to address.'[2]

This erosion is the central anthropological twist which goes along with significant current cultural developments, and which also apply to the two best known contemporary cultural as well as philosophical movements, transhumanism and critical posthumanism. Both traditions employ the word 'posthuman' within their reflections, which was the main reason for referring to the field of studies, which is concerned with the reflection on the eroded dualistic boundaries between the human and 'the other', as posthuman studies. At the same time, it needs to be noted that there is a great plurality of meanings associated to the word 'posthuman'.

> In critical posthumanism the 'posthuman' stands for a philosophical redefinition of the human. The posthuman is not the result of biological evolution, but the result of a cultural process. During the last 200 years of the unfolding Enlightenment, thinkers have criticized absolutes. Darwinism undermined humanist convictions of human exceptionalism. For posthumanists, humans are not categorically separate from the natural world but are entirely embedded in processes of becoming, so that the difference between humans and other animals becomes a gradual one.
>
> The majority of transhumanists also embrace a naturalist, nondualist understanding of human beings. However, the transhumanist focus is on the normative implications of this revised understanding of human beings, promoting our capacity to continue our animal-to-human evolution on to a posthuman stage, using pharmaceuticals, genetic engineering, prosthetics, and eventually mind uploading. For

> transhumanists, the posthuman, the evolved human, might still consider itself 'human,' or it might begin to consider itself as belonging to myriad new species. Even though both posthumanists and transhumanists use the notion of the 'posthuman' to discuss implications of emerging technologies, there is little intellectual exchange between members of these traditions.[3] (Sorgner 2017)

It might even be legitimate to highlight that there is a widely shared hostility between members of both traditions, besides that fact that both traditions embrace a new anthropology and reject the traditional humanist account of a human being as consisting of an immaterial soul and a material body.

> Transhumanists use this new anthropology as a warrant for a grand vision of posthuman progress, many posthumanists suspect the transhumanists, and in particular the goal of mind uploading, of being a crypto-religious affirmation of spirit–body dualism and human exceptionalism. Are transhumanists simply hyper-humanists? Transhumanists insist, in response, that mind uploading is a scientific project, not a supernatural one; the goal is copying the brain's wetware, not capturing the soul. As some posthumanists affirm a neo-Spinozist ontology, it can even be argued that for them it should not be problematic to attribute mental capacities to a computer.[4] (Sorgner 2017)

However, most critical posthumanists are extremely critical of mind uploading, as they identify this procedure with a traditionally humanist endeavour. They suspect that it presumes a dualist anthropology. This is a misunderstanding of the concept of mind uploading, as it is a technological project involving an empirically accessible mind and body, and not a hyper-natural one, which presumes an immaterial mind.

Human exceptionalism and its Greco-Roman Pedigree

Being intellectually engaged with philosophical challenges concerning the variety of meanings of the concept 'posthuman' is a central aspect of doing posthuman studies. There is a wide range of further challenges which are founded upon the posthuman paradigm shift, as the history of human exceptionalism can be traced back for more than 2,000 years and has had an enormous impact on our current cultural condition both in an explicit manner as well as due to it having left encrusted cultural relicts. The history of human exceptionalism is closely connected to Rome, even though initial traces can already be found in the Platonic context of Athens. Human exceptionalism, which is rooted in the understanding that humans have a categorically special ontological status in the world, has been justified by reference to the immaterial human soul, free will, rationality, as well as humans having been created in God's image, which is the reason for us having the divine spark in us. Thereby, human exceptionalism is connected to dualism, anthropocentrism, and essentialism. Only human beings possess the immaterial divine spark. Other natural entities belong solely to the natural world. This self-understanding creates a duality between the solely natural beings and human beings, whose nature is dualistic, natural as well as non-natural by possessing an immaterial divine spark. In the natural world, only human beings possess the divine spark, which makes human beings special. This logic has justified an anthropocentric bias. Furthermore, the natural world is understood as world of permanent change, whereas it is the non-natural world which is characterized by an essential stability. Consequently, it is this self-understanding which justifies an essentialism. Thus, human exceptionalism is connected to substance dualism, anthropocentrism, and essentialism.

Furthermore, human exceptionalism has gained further cultural connotations. Our exceptionality lies in our divine rationality. Men are more rational than women. Thus, sexism was generated. The divine was connected to the sun, whereas the earth was related to darkness. Thus, racism came about. Humans possess a divine spark, whereas this does not apply to other animals. Thus, speciesism was justified. Truth in correspondence to the world can only be grasped by means of our immaterial rationality. Thus, aletheism gained relevance. Given this short summary of the various implications of human exceptionalism, the logic of various types of discrimination can be better understood. The humanities, too, are structured around this human exceptionalism.

> The word 'humanities' comes from the Latin word *humanitas*. The use of the word *humanitas* in the Stoic context and in the writings of Cicero has been particularly influential. Here it can be identified with the Greek word *paideia* (παιδεία), which can be translated as *education*. This identification comes out clearly in Cicero's 'De Oratore' (1, 71). In addition, the Latin author Aulus Gellius from the second century CE explicitly identifies *paideia* and *humanitas* in his text 'Noctes Acticae' (XIII, 17). In this tradition, the concept *humanitas* encloses both ontological as well as ethical elements.
>
> The same can be said of the word *dignitas*, which was first used by Cicero in his book 'De officiis' in such a way that it attributes a nongradualizable quality to all human beings, or rational beings, which are the same according to Cicero, and which implies the demand that all bearers of the quality need to be considered in a special way. Cicero's notion of *dignitas* was received and further developed by Kant. His understanding is still of central relevance in German foundational law.

The concept is an ontological one, as it implies a certain anthropology. It is also an ethical one, as it demands that the entities that possess dignity ought to be treated in a specific manner. *Dignitas,* in this sense, was derived from the word *humanitas.*

The notion of *humanitas* has similar implications. First, it implies a categorically dualist anthropology, which affirms a categorically special status for human beings. This comes out clearly in Cicero's affirmation of an immortal soul. The Greek notion of *paideia* has comparable implications, which can be seen in Plato's analogy of the divided line. Second, both *humanitas* and *paideia* have an ethical dimension, which traditionally has been connected with perfectionist ideals. The Stoic sage, who possesses all virtues, and Plato's philosopher king are two such examples.[5]

Metahumanites – humanities dance the twist

The humanities not only have been derived from the Latin concept *humanitas,* but also still have analogous ontological and ethical implications. However, some of these implications are under attack in an age of transforming sciences and emerging technologies, as these developments are related to new insights that move away from a categorically dualist ontology and expand the options for educating youth. The scientific and philosophical works of Darwin, Nietzsche, and Freud have been particularly relevant for this development. More recently epigenetics, embodied mind theories, meta-, post-, and transhumanism, and evolutionary psychology, aesthetics, and economics have promoted this move away from the traditional humanities. A clear awareness of the similarities and differences between meta-, post-, and transhumanisms is essential for having a theoretical framework for analysing contemporary issues. It also needs to be highlighted that meta-, post-, as well as transhumanism are not monolithic concepts. Each one of these approaches can come about in several versions, for example, transhumanism show up in at least four different variants: (1) foundational transhumanism, (2) original transhumanism, (3) classic transhumanism, and (4) Euro-Transhumanism. Incorporating these new theories into the humanities leads to what I call the *metahumanities.* I am using the prefix *meta,* as meta means both *beyond* as well as *in between.* Metahumanities lie beyond the humanities, but not in such a way that old insights are being abandoned, but in such a way that former cornerstones are being integrated in an appropriate way. Hence, metahumanities also lie *in between* the humanities and new approaches such as posthuman studies. Consequently, the metahumanities can be understood as an inclusive approach. They consider a traditional version of the humanities, but at the same time they grant appropriate recognition to a new way of dealing with the world within posthuman studies.

The metahumanities embrace, integrate, and consider various emerging theories in their regular humanities curriculum, which adequately accommodate theories of evolution, and which are non-dualist, non-anthropocentric, and non-essentialist, for example, evolutionary epistemology, aesthetics, psychology, ethics, and economics; embodied theories of the mind; epigenetics; new types of spirituality and mysticism; non-dualist accounts of rights and dignity; revised concepts of the family; immanent concepts of the good life; big history; the relevance of cultural history with respect to norms, values, and the lie; and non-anthropocentric ontologies and the avoidance of speciesism.

As their insights are being widely shared, they ought to be integrated into the traditional humanities curriculum as additional topics, so that the metahumanities can come about. This idea does not imply that the new theories ought to replace the older ones, but merely that an extension of the traditional curriculum should take place, so that pupils get a more inclusive education.

Metahumanites realize the twist of humanities with posthuman studies

> To explain the relevance of reflections from posthuman studies, I have already analyzed elsewhere how this can be done in the case of a specific example in the field of music drama … With each emerging technology, new philosophical challenges arise, and with it the need to deal with them within posthuman studies.[6] (Sorgner 2017)

Posthuman studies employ the emerging theoretical terminology which has been developed by trans-, post-, and metahumanists in order to analyse philosophical, social and cultural phenomena. The relevance of posthuman studies has increased significantly in the past decade.

> In 2014, the 3rd World Humanities Forum, which took place in Daejeon City, and which was organized by the UNESCO, the Ministry of Education, the Republic of Korea, and the Executive Committee of the Organizing Committee, was dedicated to the topic 'technology' to reflect upon how emerging technologies can be considered within the humanities. In 2017, the 'Journal of Posthuman Studies', the first academic journal explicitly dedicated to the posthuman, was established. In 2019, the 8th Winter School of the Estonian Graduate School of Culture Studies and Arts will be concerned with the topic 'The Humanities and Posthumanities: New Ways of Being Human'. I have been involved in all these events. It is fascinating for me to understand how the consideration of posthuman reflections in the

context of the humanities has developed in the past couple of years. It moved from rather theoretical reflections to a dynamic area in which a great amount of complex research occurs. The enormous variety of doctoral studies as well as post-doctoral projects which are currently being researched is amazing indeed.

Even in high schools, pupils deal with posthuman issues such as the philosophy of the series 'Black Mirror'. My own students, too, are fascinated by the course 'Posthuman Studies', which I established at John Cabot University in Rome in spring 2016. Yet, there is still some debate concerning the appropriate terminology to refer to clearly distinguish metahumanites, posthumanities, and posthuman studies. In a Winter School in Tallinn which took place in 2019, they employed the word posthumanities. I find the word slightly misleading. Instead of 'posthumanities', I prefer to use the term 'metahumanities', as the prefix 'post' seems to abandon the past, whereas the prefix 'meta' means both 'beyond' as well as 'in between', which stresses both that it is moving beyond the traditional foundations and scope of the humanities, but does so in an inclusive manner, as it is in 'between' the traditional humanities, and the emerging field of 'Posthuman Studies'.

It is important to stress that since the time of their initial development by Varro, Cicero and others, the humanities have strongly been influenced by dualistic ontological thinking, that is, there is a material body and an immaterial soul or later that there is an object and a subject. What is clear, however, is that the traditional dualistic ontological framework which was dominant in the humanities for 2000 years is no longer suitable for successfully dealing with our contemporary challenges, as our anthropological outlook has changed. Post-Darwin more and more people have moved away from the understanding that we have an immaterial soul and a material body or that the subject-object-distinction is the appropriate way to classify entities in this world. Critical post- as well as transhumanists agree in this respect. Scholars, artists, and activists of both traditions develop their specific ways of suggesting alternatives for our future. The future is open, and our students as well as we ourselves need to face the challenges which go along with the new ways of conceptualizing the world as well as with the emergent technologies which enable us to shape our future. Posthuman studies are concerned with all these questions. This is the reason why we must desperately revise our traditional dualistic ontological foundation of the humanities. We have to be more inclusive and consider the challenges which occur in posthuman studies. This is what the metahumanities attempt to achieve. Metahumanities do not abandon the past, as the future needs the past. We need to study our influential and long established dualistic ontological traditions,

> which have taken us to where we are right now. However, we also need to be aware of the latest technological developments and critical post- and transhumanism in particular ought to be considered for a comprehensive conceptual understanding of the world we live in. This is what is being done in the emergent field of posthuman studies. Here, reflections concerning the impact of emergent technologies in the lifeworld are being dealt with. However, to being able to grasp all the implications of emergent technologies, a direct awareness of what digital data can do and how it can be applied is essential, too, that is, digital literacy is needed[7]. Having acquired digital literacy, the possibilities for investigations in the humanities also gets altered. Using digital technologies within the humanities becomes an option. Consequently, digital humanities need to be considered as an additional methodology within the humanities. In the age of the fourth industrial revolution, humanities students need to be able reflect upon the impacts of emergent technologies (Posthuman Studies), need to know what digital technologies can do (Digital Literacy), and need to be able to include new technologies within their research (Digital Humanities). The inclusion of all these disciplines into the humanities can be referred to as metahumanities, as it goes beyond the humanities, but it is also inclusive concerning the reflections on the latest emerging developments.[8] (Sorgner 2018b)

My own engagement has focused on the development of the field of posthuman studies.

> 'Posthuman Studies' is the engagement of the liberal arts with the impact of emerging technologies on the basis of trans- and critical posthumanism. It is different from the 'posthumanities', which have been defined as follows: 'Posthumanities situates itself at a crossroads: at the intersection of 'the humanities' in its current academic configuration and the challenges it faces from 'posthumanism' to move beyond its standard parameters and practices'.[9] Both of these disciplines are located primarily within the humanities. However, the posthumanities merely considers challenges from posthumanism, whereas posthuman studies is more inclusive by covering the wide range of non-dualistic emerging philosophical movements, whereby critical post- and transhumanism are the most dominant intellectual approaches in this respect, but there are many other movements which get considered, too, like metahumanism or Afrofuturism.
>
> The 'Digital Humanities', on the other hand, is a discipline which can be associated with many departments (philosophy, history, literature, art history ...). It can also represent a separate

department within the humanities or in between the humanities and the computer sciences. Two aspects which are usually covered in the digital humanities are research on 'digital culture and society', and the exploration of the 'use of advanced technology-related methods in humanities research', which is usually the main feature of this academic discipline.[10] As the name says, it is a discipline belonging to the humanities, which, however, employs and deals with methodologies which traditionally have been used primarily in the social, and natural sciences or mathematics. In order to properly engage with the digital humanities, a solidly grounded digital literacy is needed. Digital literacy becomes a new prerequisite for these kinds of studies. By the way, it is not necessary to employ digital humanities within posthuman studies or posthumanities, as both disciplines are rooted within the humanities. Yet, digital humanities are being considered and used within both academic disciplines. In the same way, as digital humanities slowly receive additional consideration within more and more humanities departments all over the world. Awareness of the latest technologies as well as the capacity to use technologies becomes permanently more important for experts in the humanities. This does not mean, however, that the skills, which one can learn in the humanities, become less important in an age which is characterized by an increase of digitalization, automation, and other emerging technologies. The capacities to think outside of the box and to be creative are central skills in an era of digital quantification. [This is, at least, the way, what many scholars used to think. However, there is the need to go one step further: 'Thinking outside the box only proves the box still rules you. The future begins when we build new boxes.'] These skills are not helpful solely for humanities scholars, but also for natural scientists. Hence, it is advisable to also make certain humanities courses obligatory for natural scientists. Thus, natural scientists develop further their [skill to build new boxes, or at least to realize the relevance of the humanities scholars who possess the capacity of doing so]. Natural scientists need to learn to embrace words, in the same way as humanities experts need to learn to embrace numbers.[11] This is what a twisting of traditional disciplines implies. It is not just the case that natural scientists are taking over the humanities departments. Analogous movements need to occur within the natural sciences.

Furthermore, there are additional fields of academic studies which are concerned with the impact of technologies. In the social sciences, technologies are primarily being investigated on an empirical basis. Here, one finds 'Science Technology Studies' (STS), which is concerned with 'science and technology as socio-material phenomena, including

> their historical and contemporary production and their associated forms of knowledge, expertise, social organization and controversy'.[12]

STS needs to be distinguished clearly from 'Future Studies' aka 'Futurology', which, according to the 'Oxford Dictionary of Philosophy' can be defined as follows: 'The activity of predicting the state of the world at some future time, by extrapolating from present trends. Mainly a pseudo-science, given the complexities of social, political, economic, technological, and natural factors'[13] (Blackburn 2016, 145). Neither STS nor posthuman studies or the posthumanities have anything to do with futurology.

> Even though, posthuman studies is an inclusive discipline, it is also relevant to keep in mind that this does not mean that there are no limitations, and that 'anything goes'. The developments in the realm of posthuman philosophies are extremely dynamic. The standards are not yet as established as they are in other disciplines. Various approaches challenge the intellectual grounding on which research used to take place. This makes posthuman studies a vibrant, exciting, and challenging academic field to deal with. However, this is also the reason why it attracts some self-styled prophets. By reflecting upon the concept of the discipline, and the limits of scholarly academic reflections, it becomes possible to separate solid scholarship on posthuman issues from ungrounded prophecies. It is intellectually exciting to realize how the quality of scholarship has increased in the past ten years. It was in 2009, when I had organized the first Beyond Humanism Conference which took place in Belgrade. I am already curious to see how the dynamic field of posthuman studies will develop in the next ten years, and it is fascinating to being part of such a vibrant community of scholars of the posthuman.[14] (Sorgner 2019b)

Metahumanites neither do futurology nor STS

It is important to highlight some misunderstandings when one is dealing with emerging technologies from a metahumanites perspective. Doing metahumanities does not aim for predictions of what the world will be like in 50 or 100 years. Such predictions belong to the realm of futurology and do not have anything in common with proper scientific endeavours. Science technology studies, on the other hand, are a scientific enterprise. However, they do not belong to the field of metahumanities either, as STS deals with the impact of technologies primarily from an empirical basis and thus belongs to the field of social sciences. This does not mean that empirical studies are of no relevance whatsoever in the metahumanities. However, the prime relevance is not solely being attributed to empirical studies in the metahumanities. The cultural, intellectual, and philosophical background of empirical data always needs to be dealt with and reflected upon. This is the prime difference between STS and posthuman studies. STS focusses primarily on empirical data, but when it comes to empirical data, posthuman studies always grants as much relevance to the theoretical intellection foundation as to the data which is being considered. Concerning the intellectual foundation, the philosophies of trans-, critical post-, and metahumanism are particularly noteworthy. However, it is not sufficient to merely integrate posthuman studies into the humanities for bringing about posthuman studies. Additional disciplines also need to be granted the appropriate relevance. Besides posthuman studies, the digital humanities are central, which uses technological means for investigating issues which have been of intellectual concern in the humanities.

The decisive innovation concerning digital humanities is that new technologies, particularly digitization, enable scholars to generate insights which could not have been realized without these. Which words have been the dominant in Kerouac's writings during the various phases of his career? Which words have been popular during which period of time in Japan and what does this say about the dominant cultural concerns? Is there a correlation between the quantity a word is used and the relevance concerning an author's goals in various countries all over the world? In many of these cases, digital versions of texts and data analysis skills are needed. If this is the case, humanities scholars also need to develop new skills, for example, digital literacy. I am not claiming that all the research in the humanities or metahumanities have to be founded on complex digital skills. However, in the same way as a scholar needs to be able to write, surf the web, and use the computer, a more complex digital literacy is becoming relevant for studying metahumanities. In a liberal arts university, a separate course teaching digital literacy should be mandatory as a general proficiency requirement for all students. In the same way as students need to be able to deal with numbers and words, students need to feel comfortable in the digital world. Hence, a course on digital literacy should become a general proficiency requirement.

This does not mean, by the way, that digital literacy is part of the metahumanities. It would be more appropriate to regard it as a prerequisite for successfully being able to study metahumanities. The metahumanities consist of the humanites but are being supplemented by digital humanities, posthuman studies, as well as various other emerging theories, which adequately accommodate theories of evolution, and which are non-dualist, non-anthropocentric, and non-essentialist. As a prerequisite for studying metahumanities, students need to be able to write and calculate, but they also need to have taken a course on digital literacy, and ethics, as mandatory general core requirements which are needed for all different majors. By taking a digital literacy course, students who primarily use words get a more comprehensive grounding for their research. By taking an ethics course, students who primarily deal with numbers get a more comprehensive understanding of their fields of studies.

The current paradigm shift is not a one-way street in which natural scientists take over the humanities. However, in the same way as metahumanities students need to become familiar with digital technologies, natural science students need to develop a sensitivity for the impact of their endeavours. Ethics and applied ethics provide natural sciences students with a helpful preparation for their careers as well as a better self-understanding with what they are concerned with.

After taking general core requirement courses, courses on posthuman studies and digital humanities ought to be offered. However, in addition to offering posthuman studies and digital humanities, various other emerging theories, which adequately accommodate theories of evolution, and which are non-dualist, non-anthropocentric, and non-essentialist, also need to be integrated in their courses. What does this mean in practical terms? No matter which course is being taught in the metahumanities, a broad overview of available theories needs to be accommodated. If an ethics course is taught, it is no longer sufficient to merely consider deontological, utilitarian, and virtue ethical approaches, but more critical theories which regard all values and norms as fictions also ought to be considered. If a course on epistemology gets taught, it is not sufficient to merely deal with theories of divine intuitions, British empiricism, or continental rationalism, but evolutionary epistemologies also ought to be dealt with. A course on philosophies of art must not solely consider theories which claim that we desire art for art's sake, that art ought to reveal the form of the good, or that contemporary works of art must not be beautiful, but it also ought to investigate beauty from an evolutionary perspective, reflect upon the possibility of genetically engineered organisms as artworks, and highlight the absurdity of humans who create beautiful forms, while they are embedded in a continuous fight for survival. Hence, each course ought to incorporate both historically established theories but also some cutting-edge reflections which clearly move beyond the established prejudices of traditional theories. Hence, some rather radical changes are needed to transform the humanities into the metahumanities.

Posthuman studies course

At John Cabot University in Rome, I developed a course entitled 'Posthuman Studies' in 2016. Ewha Womans University in Seoul has recently launched an master's and doctoral programme in 'Posthuman Studies'. In spring 2021, the *Posthuman Studies Reader*, which was edited by Evi Sampanikou and Jan Stasienko, was published. It includes seminal text from the wider range of beyond humanism approaches, and which have been divided up into transhumanism, critical posthumanism, and metahumanism. A pedigree of the intellectual ancestors and the development towards the posthuman paradigm shift has also been presented as part of the reader.[15] The various institutional endeavours in the field of posthuman studies make it clear that the great variety of beyond humanism philosophies represent a helpful intellectual tool for approaching the various posthuman challenges. They enable students, scholars, and intellectuals to engage with, face, and deal with the wide-range of permanently new challenges which arise in this field, and which have to do with all aspects of the lifeworld from religion, via politics, and economics, to the wide range of artistic disciplines.

Conclusion

The posthuman paradigm shift has implications for the structuring of higher education. In addition to students taking mathematics and language courses, new skills become essential, in particular digital skills, and a sensitivity for ethical challenges, as new technologies imply new moral challenges. Hence, it seems necessary to expand the list of courses which belong to the general core requirements. A digital literacy and an ethics course seem to be relevant for preparing students with the challenges of our times. Yet, the general core requirement courses are merely meant to serve as preparation for being able to deal with more specialized skills.

Concerning the more specialized skills in the field of the humanities, three suggestions were dealt with here, such that the humanities can be transformed into the metahumanities: (1) a course which deals with the meaning, and the impact of emerging technologies, and which provides students with the theoretical grounding for verbally addressing, and dealing with these challenges, that is, a posthuman studies course; (2) a digital humanities course which enables students to develop digital skills for handling research questions which so far could not be deal with, as digital technologies represent a pre-requisite for generating the data necessary for corresponding reflections; (3) all professors in the metahumanities ought to be encouraged to expand the range of topics which is being covered in a course, so that students are also being confronted with the latest challenges in a field of studies and not merely with the dominant approaches, reflections, and theories of the past.

All these reflections concerning the metahumanities are meant as an initial stimulus for reflecting on education in an age of the posthuman. These thoughts are not meant to represent a final word on the concept of the metahumanities, but they should merely serve as a contingent suggestion.

The future needs the past, and 'war is the father of all things'. Both insights also apply to the question of higher education in an age of the posthuman paradigm shift. In the following three chapters, I will deal with three cases which are of central relevance for the metahumanities.

The Case of In Vitro Meat

In December 2023, the Italian government implemented a ban on cultivated meats, restricting both the production and marketing of these products.[1,2] ' "We are concerned with protecting our tradition, our products and, above all, the health of Italians," said Lega politician Davide Bergamini.'[3]

> The bill sought to address these problems, stating that 'synthetic food represents a dangerous means of destroying every link with natural food and different lands by cancelling every cultural distinction, often thousands of years old'. Italy's Agricultural Minister, Francesco Lollobrigida, declared that this legislation has made Italy 'the world's first country safe from the social and economic risks of synthetic food'.[4]

Tradition, economics, nature, and health have been the decisive factors for restricting production and marketing of in vitro meat in Italy. It is the first European country to do so. In Singapore, such meat has been sold since 2020, and in 2023 the USA also approved selling such meat.[5] According to the German Verbraucherzentrale, in vitro meat gets produced as follows:

> Laboratory meat (in vitro meat) is produced by means of 'tissue engineering':
>
> In this process, muscle tissue is first removed from an animal.
>
> Stem cells are extracted from this tissue and multiplied in a container (bioreactor) using a culture medium that provides optimal conditions.
>
> The cells pass through various stages and develop muscles. The cells grow together into a larger mass via a carrier scaffold, usually made of animal collagen.[6]

Thereby, they highlighted the following issues:

> Clean meat, which is produced in the laboratory from cultured cell cultures, is considered a novel food and requires approval in the European Union (EU).

> Whether the production of lab-grown meat is more environmentally friendly than conventional meat production cannot yet be clearly proven at this early stage. It could save resources such as land, but currently requires more energy.
>
> Animals are still usually killed for the nutrient solutions used to grow the cells. This contradicts the idea that many consumers have of production without animal suffering.
>
> Transparent labeling and comprehensible information are necessary. The general benefits, harmlessness and health value must be made clear.[7]

One of the issues raised needs to be relativized. It concerns the nutrient solution, whereby a foetal calf serum was used as a culture medium. This no longer has to be the case, as 'companies such as Mosa Meat state that they produce without this substance. In this way, animal suffering can be avoided.'[8]

Peter Singer's reflections have made more people become vegetarians than that of any other philosopher, as he so forcefully described the relevance of avoiding suffering. He clearly stated the following in an article in *The Guardian* from 2013: 'I haven't eaten meat for 40 years, but if in vitro meat becomes commercially available, I will be pleased to try it.'[9] He stuck to what he said. In a post on Twitter from 7 September 2023, he highlighted:

> After 52 years, I have eaten chicken again! But no animal was raised and killed for my meal. It was cultivated chicken, grown from cells by
> @GOODMeat
> - yes, that's correct, google it - and Singapore was, until very recently, the only place in the world where you could eat it.

Thereby, he posted a photo together with his former doctoral student Julian Savulescu, who has moved from Oxford to Singapore and who invited Peter Singer for some lectures there after he had retired from Princeton. This is the situation with which we are confronted nowadays.

Counterarguments

Despite all these developments, and even though a foetal calf serum is no longer needed to produce in vitro meat, the Italian government has forbidden the production and marketing of these products. Their main arguments were tradition, economics, nature, and health. Are these good arguments?

What about the issue of tradition? Traditions do give people stability. In a world of permanent change, I am aware of the relevance of such stabilities. Having a home, a place one belongs to. These are widely shared longings, in particular among those who live in a certain area during all stages of their lives, and it needs to be highlighted that this applies to an enormous number of human beings. If you just consider the issue of tradition, the enormous number of challenging issues become obvious. Female genital mutilation has been a widely shared tradition in many cultures. Same-sex relationships have been criminalized in many regions of the world. Many communities believed that witches deserve to be burned. Just because something is a tradition does not make it praiseworthy or worthy of being conserved. The issue needs to be reflected upon, and whether it is morally appropriate or not. The three examples I mentioned clearly highlight that it is praiseworthy that some traditions have been abandoned. Is it good to raise animals in animal factories until they can be killed for human consumption, or would it be better to have meat without having to do all this?

What about the issue of economics? This point is interesting to consider, as the production of so-called natural food and drinks is particularly relevant for the Italian economy. Many family businesses, individuals, and craftsmen financially depend upon the natural goods they are producing. It is not so easy to change the infrastructure, logistics, and available skills of in an economic ecosystem. In the Italian ecosystem, the production of so-called natural food and drinks is particularly relevant. The Italian government might indeed have made a strong move to protect the producers of these goods by implementing the bill concerning in vitro meat. However, if in vitro meat becomes particularly popular, then the introduction of this bill merely postponed the downfall of these producers for a while. This thought, too, needs to be considered, when reflecting upon the issue whether the bill is in the long-term interest of Italy. Only by considering the long-term economic developments related to in vitro meat, can this issue be clarified further. It might be relevant to consider that Nokia was at the peak of its success in 2007, when the iPhone was launched. German car manufacturers have long focused on combustion engines, while the Chinese car producers and Tesla had already been working on electric cars.

What about the issue of nature? There is a widely shared yearning for natural products. A sustainable future in a natural environment is a goal desired by many people. Cows should live in the Alps, where a natural environment still exists. However, it does not need much thinking to realize that nature has become unnatural with the coming about of human beings. In a natural world, antibiotics, anaesthetics, and vaccinations do not exist, neither do houses, central heating, or tap water. Earthquakes are natural, the coming about of humans with a severe mental handicap is natural, and malaria is natural, too. To identify nature with goodness, and culture with something bad is meaningless. Whether in vitro meat is good or bad does not depend upon the question of whether it is natural or not.

What about the issue of health? This is indeed a great question, as most humans do indeed identify a prolonged healthspan with an increased quality of life. Further studies are needed to clarify all the health implications of in vitro meat. However, what needs to be highlighted already is that in contrast to regular meat, it gets produced in a clean environment. Animals which are raised for consumption need to be given antibiotics, which again increases the likelihood of the coming about of antibiotic-resistant cells. Of course, unless chicken or fish get produced, what is realized is still red meat, and there seems to be a correlation between the consumption of red meat and colon cancer.[10] However, this issue applies both to regular as well as to in vitro meat. Besides this issue, it seems to be the case that in vitro meat is healthier than regular meat. Furthermore, additional nutrients could be added to make it even healthier. Yet, the health implications of in vitro meat need to be studied carefully. So far, there is no strong indication that in vitro meat is unhealthier than regular meat.[11] Given the plausibility of the previous reflections, the main counterarguments presented by the Italian government do not apply necessarily.

Additional reflections

What else can be said concerning the practice of producing, marketing, and consuming in vitro meat? Social, cultural, and religious aspects of meat consumption always need to be considered, too. How in vitro meat is seen in society also depends upon the phrases being used to talk about it, and the emotional connotations of these words. There was a study that investigated how different terms alter the reaction to this phenomenon in Italy.[12] The study slightly revised the hierarchy of terms compared to a similar study conducted in the UK a few years earlier, which classified the terms as follows.[13]

The term 'clean meat' sounds particularly appealing. Cleanliness is associated with health, flourishment, and excellence. Slightly less well connotated was the term 'cultured meat'. However, culture in general is a word which has rather positive connotations, so cultured meat must always be something positive. Concerning its popularity, 'in vitro meat' was behind the other two concepts. It can be assumed, that the lab is not an environment which people like to associate with eating and drinking and other sensual pleasures. The least popular term was that of 'artificial meat'. The word artificial paints a picture which is opposed to that of happy cows on an Alpine meadow.

The new study from Italy highlights that today, 'cultured meat' is the dominant term in traditional media and formal discourse (for example, academic papers, policy reports, news reports). The concept of 'clean meat' is more associated with advocacy, non-governmental organization (NGO) campaigns, and emotional communication, especially on social media. The thesis explicitly states that 'cultured meat' is more neutral and widely adopted in institutional contexts. In contrast to previous studies, the term ranks higher in usage and perceived neutrality than 'clean meat'. The order of the other two terms remains unchanged.

Consequently, when talking about the topic, the appropriate choice of words needs to be reflected upon. It can be compared to the use of liberal eugenics or genetic enhancements. Even though the same procedures are being referred to, eugenics alludes to terrible processes which occurred during the existence of the so-called Third Reich. Enhancement, however, is a term with enormously positive associations. Who does not want to be better off?

This striving to be better off also resonates with another dimension of meat consumption, the relevance of social distinction. For a long time, meat consumption has been identified with being rich. If you were lucky, it was possible to eat meat maybe once a week, maybe on a Sunday for lunch when the entire family comes together. Given the economic growth in China, it is no surprise that meat consumption has increased significantly there, too. However, the religious dimension of meat must not be underestimated either. Given the predominant Buddhist outlook in India, meat consumption is significantly lower there than in a country like China. These considerations reveal which elements need to be considered when reflecting upon meat and in vitro meat, and its possible futures.

A further tendency needs to be noted in Western countries. There has been a cultural posthuman paradigm shift towards non-dualistic ontologies of becoming after Darwin, Nietzsche, and Spinoza (Sorgner 2010c). Consequently, it is no longer widely shared that human beings are the coronation of creation, but, according to many people, human beings are one species among many. We are completely entangled with the rest of the world, and not categorically but merely gradually different from it. Consequently, an alternative grounding for our moral status needs to be reflected upon. In addition, by mindfully reflecting upon our embeddedness, the entanglements of all our actions get realized (Sorgner 2022b). This type of mindfulness meditation can be applied to any of our encounters, also meat consumption. Thereby an awareness gets realized concerning suffering due to the overall conditions in animal factories, the space animals need to take up, the cruelty of getting killed in such a factory, how soil gets polluted and carbon dioxide gets emitted due to such factories, which has enormous environmental consequences, the worrying health conditions in such factories which are the reasons why antibiotics need to be given all the time, which again increases the likelihood of the coming about of antibiotic-resistant cells, and this has serious implications for human health. Meat transportation further increases global warming, the likelihood of colon cancer gets increased by eating red meat. This line of thought could be continued much further. However, it already reveals how intimately entangled we are with the world such that the social, economic, political, ethical, and environmental implications of meat need to be reflected upon. Such reflections can also be done with respect to in vitro meat, whereby it becomes clear that many of the challenging aspects of traditional meat production get undermined. However, I suspect that the moral issues with in vitro meat are not so much related to specific intellectual worries but rather to a general cultural shift which has been taking place for quite some time.

From nature to culture

There are many philosophies of nature, but a particularly influential one in the Western tradition has been the Aristotelian natural law tradition which was developed further and integrated into the Catholic culture. Thus, this way of thinking has become dominant in the Western world (Sorgner 2022b). In this way, a natural purpose has been associated with all aspects of our ways of life. It has also had consequences on our understanding of the concept of health. If our body fulfils its natural functions properly, it is healthy. If this is not the case, it is ill. If the eye sees, it fulfils its natural function. If the nose smells, it does so, too. If the genitals reproduce, they act according to their nature, and thus, are healthy. Consequently, masturbation, using contraceptives and same-sex relationships are unnatural. Homosexual acts have long been seen as not only unnatural but even as a disease as well as acts which need to be penalized. Thus, it has been argued for a long time. Up until 60 years ago, homosexual acts have long been seen as acts which need to be penalized by law in many Western countries. Transgender issues and sex-assignment surgeries are still widely being discussed both from a legal, cultural, as well as from a moral perspective.

The decisive shift which has taken place is that from nature to culture. Sexuality was only seen as morally legitimate in Catholic natural law tradition, if it was aimed at procreation. The condom became available, then the pill, in vitro fertilization, surrogate motherhood, children with three biological parents, whereby further cultural options for procreation have become available. Artificial wombs are available for lambs; they are also being worked on for human beings. Biologically related offspring is possible for two male mice.[14] Scientists still try to make it available for men. The same applies to bioprinting, which eventually might provide human beings with an entirely new option for realizing a new generation.

The catholic natural law gets challenged, while more and more cultural options become available which are more reliable, bring about better consequences, and offer additional options. However, there is a widely shared status quo bias, which can probably be explained by a fear of the future, a fear of the unknown (Bostrom and Ord 2006, 656–79). Things are working as they are. Why do we need to introduce radical changes? Innovations might lead to the end of humanity. Any technological innovation in the past has led people to raise such fears. 'For example, many Native American tribes believe that photographs can steal a part of a person's soul. This belief is connected to their spiritual practices and their view of life's interconnectedness. Historically, some tribes thought that a camera's flash could disrupt a person's spirit or attract evil spirits.'[15] It did not happen, and the same applies to many other instances of technological innovations.

There is a widely shared goal that by moving beyond traditional practices, a fully technologically controlled institution comes about, which many see as a worrying dystopia. However, at the same time we need to realize that we hardly manage to control anything. Earthquakes, pandemics, volcano eruptions, asteroid impacts, cancer, Alzheimer's disease, Parkinson's disease, amyotrophic lateral sclerosis, strokes, heart attacks … These things occur. Tragic events happen every day. Luckily, we have managed to develop many technologies, which enable us to live longer healthily. A realistic goal cannot be to control all of nature but to find as-good-as-it-gets solutions to some of the most problematic challenges. Luckily, we have already managed to find many of them already. Anaesthetics, antibiotics, and vaccinations have had an enormously beneficial effect on the qualities of our lives. More and more people live longer healthily, and this does not only apply to people in socially advanced countries, it applies to countries all over the world.

As part of the posthuman paradigm shift, we have moved from a categorically dualistic understanding of human beings concerning their ontological essence towards a non-dualistic version (Sorgner 2022a). This implies that we are merely gradually different from all the other entities around us, and no longer categorically different. Consequently, we also try to reduce the suffering of non-human animals. This goes along with the posthuman paradigm shift. There is an enormous amount of suffering connected with animal factories which are the basis for producing regular meat. In addition, these factories also promote global warming and have further challenging environmental implications. Moreover, the antibiotics which need to be given permanently lead to the coming about of highly dangerous antibiotic-resistant cells. It seems to be the case that many of these ethical issues can successfully be dealt with by means of in vitro meat. Of course, many pragmatic challenges need to be dealt with still, like that of scalability. However, in order to see whether they can be dealt with successfully, in vitro meat needs to be produced and offered. It seems that there is a demand for this kind of meat. In vitro meat seems to be an as-good-as-it-gets solution when environmental, ethical, as well as pragmatic issues get considered.

Conclusion

Given the previous reflections on the morality of in vitro meat, it seems to be the case that from an ethical perspective, there are far more good arguments in favour of producing and marketing in vitro meat than there are against doing so. The most serious ethical worry might be that of a potential health issue; however, there is no solid reason for why this ought to be the case. Besides this issue, the reasons for banning the production and marketing of in vitro meat rather seem to be associated with a general fear of technological solutions, of the future, and of the unknown in general. If we just look back the past 200 years, we need to realize that the human life expectancy has doubled in socially advanced countries, and even the socially least developed countries need to acknowledge a significant increase of human life expectancy in their countries.[16] It also needs to be noted that the global absolute poverty rate has dropped from over 90 per cent 200 years ago to about 10 per cent nowadays.[17] Given all these developments, I wish to highlight that I can hardly wait for our posthuman future to occur. For now, in vitro meat seems to be an as-good-as-it-gets solution to many ethical issues which go along with traditional meat.

The Case of Surrogacy

CNN reported the following on 19 October 2024:

> Italy has broadened its legislation on surrogacy, which has been illegal in the country since 2004, to now criminalize 'surrogacy tourism' in countries like the United States and Canada, subjecting any intended parent who breaks the law to fines of up to €1 million ($1 million) and jail terms of up to two years.[1]

The Italian prime minister Meloni invoked a Kantian principle to justify the law, consonant with the principles underlying the German legal system: 'Meloni welcomed the Senate's decision on X Wednesday, calling it "a commonsense rule against the commodification of the female body and children. Human life has no price and is not a commodity."'[2]

The body is not just a human body, but it is covered by human dignity. According to Kant, anything possessing dignity has infinite value. Only objects are regarded as property, that is, have an exchange value and can be bought and be sold, as they have a finite value. Either an entity is an object or it possesses dignity. Rational entities have dignity. Whatever has no rationality is an object. Only objects may be commodified or treated merely as a means. A person must never be treated merely as a means to an end. A central moral challenge is what these demands imply.

According to the Basic Law for the Federal Republic of Germany, a competent adult who wants to earn money as a peep show dancer, is legally not allowed to do so, as the person would allow him- or herself to get treated merely as a means.[3] What is missing in this set-up is the recognition of the other as a person due to a one-way mirror being placed in between dancer and spectator. At the same time, he or she may work as a sex worker, as it is merely a contract between competent adults concerning the use of specific skills. Sex workers do not sell their bodies. They sell a specific skill. If it were the case that their body was sold, then the buyer would have the right to amputate their limb, to kill the other, or to do whatever is in the owner's interest. This is not the case for sex work. The contract merely regulates the temporary use of a specific skill.

What happens when it comes to surrogacy? Clients and potential surrogate mothers negotiate a contract. If the contract is such that both parties, which consist of competent adults, agree to it, then the contract can be signed. How should it be interpreted that a womb gets used for many months without any break? Does the client sell her womb or merely offer a specific skill, the skill of being pregnant and giving birth to a child, for a specific period of time? If we analyse what it means to sell an object, then the buyer would also have the right to remove the womb. However, this is not the case. There are rigid limitations to what the client is allowed to do. Hence, it clearly does not seem to be the case that someone's womb is being sold. I agree with Meloni in so far as a 'Human life has no price and is not a commodity.' However, it clearly is inappropriate to identify surrogate motherhood with the commodification of human life. The skill of being pregnant and of giving birth is being sold, and the person has to work for nine months without a break. This might indeed be problematic. However, the argument of the commodification of a human life would not apply, nevertheless.

A different argument could, however, be used to argue against the legal legitimacy of surrogacy, namely that it is unnatural. Here, the notion of 'natural' must be specified. In this context the reference to nature usually refers to a functional understanding of humanity which is rooted in an Aristotelian–Thomistic understanding of the natural law. This means that only if a body part fulfils its natural end, an act is natural and hence morally legitimate. Hence, the eye needs to be used for seeing, the nose for smelling, and the genitals for reproductive purposes. Any other use is unnatural and, hence, morally illegitimate. This has significant implications. Same-sex relationships, the use of condoms, the pill, in vitro fertilization, having a child with three biological parents, surrogate motherhood, the use of an artificial womb, and eventually bioprinting are all unnatural, based on this understanding of nature, and consequently morally illegitimate. The artificial womb does not yet exist. However, scientists are working on it, and biobags have already been realized for sheep.[4] Bioprinting life is still under development. However, initial successes have to be recognized already. It will probably still take some decades until it can become a realistic option.

Anaesthetics, antibiotics, as well as vaccines are unnatural, too, given this view, whereas being born with a strong mental handicap, getting cancer, Alzheimer's disease or Parkinson's disease, and being killed by a virus can be fully natural. To identify nature with goodness and culture with badness is far from plausible. Given the enormous achievements that have been brought about by the invention of anaesthetics, antibiotics, and vaccines for the quality of human lives, it might be more plausible to argue that the move from nature to culture has increased the likelihood of persons living good lives. All these technical options definitely increase the plurality of lifestyle choices which persons can embrace, which is an enormously wonderful achievement. These developments are such that sexuality continuously gets disentangled from reproduction such that sex can be used for fulfilling special needs, and technologies for improving the quality of reproduction.

A technology for having a child with three biological parents is available already. Hence, alternative types of relationship can be realized. A polyamorous bond between one man and two women can be lived, and they can have a biologically related child together. Such children have already been brought into existence, for example, in the UK. This seems to me as fairly similar to a traditional family, father, mother, mother, and biologically related child. If the artificial womb becomes available, then the range of choices could be widened further, which might reduce the discrimination towards women, too. What happens if these three people uphold that they want to get married, as they love each other? Why should it remain forbidden? In Bolivia, three men have already been granted the right to get married, by the way.[5] Financially such a union would be in the interest of state, because if one person gets ill and is unable to continue working, then the others need to financially take care of the person, and not the government.

Conclusions

These reflections reveal that reproductive technologies go along with significant cultural challenges, which lead to tensions with encrusted relics of the past in which we are still legally embedded. I understand that these options are feared by many people, in particular those whose lifestyle is more rooted in a specific city. For digital, intellectual, and financial nomads, on the other hand, such developments may go along with enormous possibilities. Given our past experiences, it seems to me central to highlight that technologies increase the plurality of personal lifestyles, which is an enormously wonderful achievement from the perspective of Euro-Transhumanism.

The Case of AI, Art, and Creativity

Artificial intelligence (AI) has generated an enormous number of amazing achievements. Deep Blue beating chess world champion Garri Kasparow in 1997. IBM Watson winning Jeopardy in 2011. AlphaGo from Google beating the world's best GO player, Lee Sedol, in 2016. However, it has widely been taken for granted that the arts and the realm of artistic creativity are domains in which only humans can excel.

However, even here the question remains open, what does this mean? Which humans are actually capable of creating art? Do you need to have studied the fine arts to become an artist? Do you have to have some specific intellectual or manual capacities for doing so? Do you need to be aware of our art historical embeddedness to be able to be artistically creative yourself? Are there certain age requirements? Can a five-year-old child already create a work of art?

In 1761 Wolfgang Amadeus Mozart had already realized some compositions.[1] His *Andante in C* already reveal some stylistic traces typical for Mozart. However, it clearly is not a ground-breaking composition. It might merely get performed as a curiosity to demonstrate what Mozart was capable of doing at such a young age. Should it count as art?

Besides the issue of whether all human beings can realize works of art, the topic needs to be considered whether art can be realized by non-human animals. It might be relevant to distinguish between conscious animals like pigs and self-conscious animals like dolphins and Orangutans. However, even the latter do not seem to possess more developed mental capacities than a three-year-old human child. Maybe, self-consciousness is not even needed to create a work of art. In 2021 the painting *Wild and Free* was sold for £20,000. It was composed by Pigcasso – not Picasso, but Pigcasso, as the creator is a pig.[2] However, can a conscious animal like a pig create a work of art at all? Should it not be the case that human capacities are needed to realize this task. Pigcasso is making quite a substantial amount of money with his works. Is this a decisive factor for counting as an artist? However, as Pigcasso does not count as a person but as an object, Pigcasso cannot make money, because Pigcasso legally counts as an object which is owned.

In 2014, a law court recognized that a monkey counted as the originator of a photo. This issue had to be discussed as a selfie of this monkey went viral. Consequently, there was an enormous amount of financial interest connected to the photo. It was shot with the equipment of the photographer David J. Slater, who specializes in taking picture of nature and wild animals. However, he merely provided the equipment, but he did not push the trigger for the photo to being shot. Hence, he could not count as the originator of the photo. The monkey, Naruto, was responsible for doing so. However, monkeys, too, legally count as objects. An object cannot hold the copyright of the selfie, only a person can possess it. Without a legal copyright holder, the enormously successful photo became available in the public domain. Everyone was able to use it free of charge.[3]

If the legal stance is so clear in the previous two cases, should the same regulations not apply to AI, too? An AI is part of a computer, and a computer is an object which is owned by someone. However, in 2018 the *Portrait of Edmond Belamy* which had been created by an AI was sold for $432,000 at an auction. It was realized by the art collective called Obvious, which consists of Hugo Caselles-Dupré, Pierre Fautrel, and Gauthier Vernier. It is also noteworthy that the source code of the work had been created by the young artist Robbie Barrat as an open-source code. The main contributions of the art collective Obvious was the idea and their reputation.[4]

In the meantime, more and more artistic projects involving AI have been realized. Under the direction of musicologist Matthias Röder, AI realized Beethoven's *10th Symphony* in 2021. The composer Sven Helbig has created pieces of music for his own work with the help of AI. The song 'Heart on My Sleeve', released on 4 April 2023, sounded like a musical collaboration between Drake and The Weeknd but was created by an AI. At the time, AISIS released an entire album of AI-generated Oasis songs. Subsequently, numerous popular songs by long-dead artists were sung, including a particularly successful Frank Sinatra cover of the Green Day hit 'Basket Case' in July 2023. In 2024, OpenAI even introduced a programme, Sora, which can be used to create one-minute videos using short voice prompts, making it difficult for experts to determine whether they are AI-generated videos. Google's Gemini has been making AI accessible to a much larger user group since 2023.

The question becomes more and more pressing: Can an AI creatively produce a work of art? My answer is clear. It cannot do so, as an AI does not have intentionality. Hence, by itself it does not want to create a work of art. It needs input. It needs a human who sets the initial trigger. Of course, we have heard of many Silicon Valley entrepreneurs, computer scientists, and AI experts, who talk about artificial general intelligence and even superintelligence. A scholar who used to be at Oxford made his entire career on his reflections on superintelligence. However, it remains entirely unclear what superintelligence is, and given the vague descriptions which have been given so far, it is clear that we are far away from any form of artificial general intelligence (AGI) or superintelligence, as intentionality, volition, being alive, consciousness, self-consciousness, emotions as well as the experience of pain and pleasure are qualities for which we so far do not have any indication that they can be realized on a silicon basis. So far, they only exist in carbon-based organisms.

Of course, this does not mean that it is impossible for these qualities to come about in silicon-based entities, but there is no good reason for claiming that this will happen soon. The strongest argument for the possibility of it being possible for them to come about in silicon-based entities is the following. Life came about on earth on the basis of inorganic material 3.5 billion years ago. If it happened then, then this could also happen on a silicon basis. This is logically plausible. However, it needs to be noted that, so far, we do not have any indication of this actually being possible.

I would reply differently to another important question: will AI be relevant for artistic processes in the future? Absolutely! This has to be expected. The innovative force of generative AI is enormous. Whenever a prompt gets asked, the AI realizes a different response. Some responses are more impressive than others, but then it is up to the human artist to select and put together what fits best. Generative AI is enormously innovative. However, it cannot decide so far what works better than something else. This is the creative task of the artist. AI can innovate, but human beings with their intentionality, volition, emotions, and self-consciousness can be creative, whereby creativity might merely consist in the capacity ask prompts, select suitable responses, polish them, if needed, and arrange them appropriately.

Conclusion

To progress in the aforementioned manner is exactly what I have done with ChatGPT. The algorithm is familiar with my Euro-Transhumanist approach. I incorporated ChatGPT to create aphorisms which take my version of Euro-Transhumanism as well as specific insights into consideration. Then, I selected, revised, and polished the ones which resonated with me best. I included them at various places in this manuscript.

Art

Walker's and Tomlinson's interpretation of 'Ficino's Magical songs', which is the title of a chapter in Tomlinson's study on *Music in Renaissance Magic* (1993), represents contrasting interpretations of two central philosophical challenges in the philosophy of music. I am drawing particularly on Walker's book-length study on *Spiritual and Demonic Magic* (2000). These titles serve as an inspiration for the title and structure of these reflections which are entitled *From Ficino's Spiritual and Demonic Magical Songs to Posthuman Reflections on Music*. Two central philosophical challenges addressed in both of their reflections are the following: (1) What is Ficino's understanding of music? Walker argues that the highest kind of music according to Ficino are sung monodies, whereas Tomlinson stresses that Ficino has a wider understanding of music which allows the possibility of spiritual instrumental music, too. (2) Can there be spiritual, natural, and demonic songs? According to Walker, Ficino affirms a binary distinction between spiritual songs, which can lead upward to God, and demonic music, which is unable to do so, whereas Tomlinson stresses that, according Ficino, not only spiritual songs can take you upwards, but this can also happen by means of spiritual, natural, and demonic songs, as 'Ficino's demons' 'were not the intrinsically evil fallen angels of Christian tradition' (1993, 123), whereby Tomlinson explicitly stresses that he is aware that 'Ficino diverged in all these views from orthodox Christian conceptions of the relations of music and demons' (1993, 125). Hence, Tomlinson highlights that, according to Ficino, there is a kind of magic, which is 'both wholly spiritual and natural and also thoroughly demonic' (1993, 126).

To explain the meaning, relevance, and implications of these philosophical issues, I will give an outline of the Pythagorean and Neo-Platonic roots of Ficino's approach. These include fundamental ontological structures, which Ficino integrated into his philosophical account of music, but there are also other elements where he clearly moves beyond Plato's reflections on music (Sier 2010). These discussions also have a practical relevance for the history of music, reflections on the ethical power of music, and the need to twist their approach in the age of the posthuman paradigm shift, whereby the relevance of beauty from the perspective of Euro-Transhumanism comes out.

Pythagorean reflections on music

The Pythagorean musical tradition is rooted in the distinction between *musica mundana*, *musica humana*, and *musica instrumentalis*.[1] This philosophical distinction even led to scientific breakthroughs, for example in the thinking of Johannes Kepler and, in particular, in his book *Harmonice Mundi* (Haase 1998), which he regarded as his central academic achievement, even though nowadays he is best known for Kepler's three laws. It is the Pythagorean way of thinking about the world, and particularly the concept of *music mundana*, which led to Kepler's first law, according to which planets move elliptically around the sun. *Musica mundana* stands for the cosmic perspective of music. According to this tradition, such music exists, and the planets create sounds on the basis of their movements. However, if they are creating sounds, how is it that we do not actually hear the sounds of the harmony of the spheres? Pythagoreans actually dealt with this question and some replied that because of the permanent presence of these sounds, we have grown so accustomed to them that we cannot consciously hear them. However, these sounds exist according to a Pythagorean concept of the world. Nevertheless, Kepler wondered how such sounds can come about, if planets move in circles around the sun. Variation is needed for a sound to come about. Hence, the planet had to have different distances from the sun at various times. This led him to investigate the possibility of planets moving on elliptical lines around the sun, and thus he developed his first law.

Musica humana, on the other hand, represents analogous structures to the ones which can be found in planets within human beings, as a way to connect the body, spirits, and the soul. It is this connection that enables music to have an ethical consequence for human beings. Certain harmonies can promote courage, others, and boundless joy, and further harmonies are responsible for a clear intellectual order and wisdom. Harmonies and melodies are closely related to attitudes, insights, and possibilities. There has been a complex philosophical investigation of the various implications of different mathematical ratios, in particular the correlations of these ratios to basic stable attitudes as well as to ontological structures that have been carefully examined. The same applies to the use of instruments, mythologies, verbal content of songs, or whether instrumental music or sung music is superior. Is polyphonic, homophonic, or monodic music morally appropriate? All these philosophical challenges have been dealt with and have had implications for the history of music. It was taken for granted that there are, at least, structural analogies between the cosmos and its *musica mundana*, with the human moral and religious attitudes, or the *musica humana*.

Last but not least, *musica instrumentalis* should be mentioned, which refers to music that we can actually listen to, the type produced by instruments or the human voice. The same numerical ratios apply to all three spheres. The entire world is sound and can be analysed by means of mathematical ratios. Thereby the direct relevance of numerical relations and the Pythagorean musical tradition has also become clear. Investigating the powers of music is concerned with reflections on mathematical ratios and numbers. These reflections happened when music was listed as one of the four liberal arts of the *septem artes liberales*.

The seven liberal arts can be divided into two groups, the *quadrivium* and the *trivium*. The English word trivial was derived from *trivium*. It means three ways, *tre vie*, and has acquired the meaning of a basic necessity, a widely shared understanding, as in the *trivium* subjects were studied, which have to do with words everyone needs to know. In the *quadrivium*, on the other hand, numbers are the subject of concern, and this is where music is listed. Yet, dealing with music does not mean composing, performing music, but it stands for the intellectual reflection on music and its powers, which is the task of a theologian or a philosopher of music.

Numerical ratios can be found in nature. In Pythagorean thinking, the *tetractys* represented perfection and beauty. It consists of the numbers that bring about consonant intervals, 1, 2, 3, and 4. The sum of all these numbers is 10, which consequently is a perfect number. To demonstrate the perfection of these numbers, the monochord can be dealt with which has been created by Pythagoreans, which consists of a resonance box and a string. If you pluck the string, then press it down in the middle and pluck it again, the result is the interval of the octave, which is why this can also be described as a ratio of 1:2. The ratio 2:3 results in the interval of the fifth, and the ratio 3:4 in that of the fourth. In these cases, the monochord is divided into three or four equal sections and then pressed down at one point so that there are two or three sections on one side and one section on the other. When the string is plucked, a note is produced that forms a corresponding interval with the note produced by the full string, either that of the fifth or that of the fourth.

It might be interesting to note that there is a clear difference between the Pythagorean tuning system and the equal temperament we have nowadays. Nowadays, 12 (equal-tempered) fifths correspond exactly to seven (pure) octaves. Yet, there is a difference between seven (pure) octaves and 12 (pure) fifths in early Pythagorean tuning. The difference is called the Pythagorean comma, which is the difference between 12 pure fifths and seven pure octaves, which is the interval of about an eighth tone or about a quarter of a semitone.

These reflections have been developed further, for example, when it comes to the notion of beauty, where the golden mean is often invoked. The close relationship between number ratios and the golden ratio is illustrated by the Fibonacci sequence. Fibonacci, who was born in Pisa at the end of the 12th century and also died there, was one of the most important medieval mathematicians. The sequence he developed results from the following recursive law of formation: ax+2=ax+1+ax, that is, 1, 1, 2, 3, 5, 8, 13, 21,

34, 55 … Finally, Johannes Kepler (1571–1630) found that the golden ratio is approximated by first creating the quotient from two consecutive Fibonacci numbers and then a sequence of such quotients: 1/1=1; 2/1=2; 3/2=1.5; 5/3=1.6666; 8/5=1.6; 13/8= 1.625 … Here, it should be noted that even though numerical ratios are closely connected to beauty in the Pythagorean tradition, Plotinus clearly highlights that beauty must be something simple. Consequently, there are levels of beauty, an ideal beauty which is connected to perfect numerical ratios, and an even higher level of beauty connected to beauty itself, which is a simple form.

Plotinus on beauty

Plotinus presents five arguments in favour of the position of why symmetry does not represent the nature of beauty.[2]

Most of the reasons are found in the writing 'On Beauty' (I 6,1), another in the later writing 'On the Immortality of the Soul' (IV 7):

Argument 1 (I 6,1,25ff.):

If the essence of beauty consisted in symmetry, then there could be nothing simple that is beautiful on its own.

However, there are simple things (colour, light, gold …) which are beautiful.

Ergo: The essence of beauty cannot consist in symmetry.

Argument 2 (I 6,1,25ff.):

Symmetry requires multiplicity and thus has many individual parts.

If the essence of beauty consisted in symmetry and thus presupposed a multiplicity, then individual parts could not be beautiful.

All parts that are not beautiful are ugly.

If the essence of beauty consisted in symmetry, then beauty would have to be composed of ugly parts.

Beauty cannot be composed of ugly parts.

Ergo: The essence of beauty cannot consist in symmetry.

Argument 3 (I 6,1,40ff.):

If the essence of beauty consisted in symmetry, then human actions and virtues could not be considered beautiful.

However, human actions and virtues are beautiful.

Ergo: The essence of beauty cannot consist in symmetry.

Argument 4 (IV 7):

If the essence of beauty consisted in symmetry, then no living human being in which there is a low degree of symmetry could be more beautiful than a likeness in which there is a high degree of symmetry.

However, a living human being in which there is a low degree of symmetry is more beautiful than an image in which there is a high degree of symmetry.

Ergo: The essence of beauty cannot consist in symmetry.

Argument 5 (I 6,1,35–45):

If the essence of beauty consists in symmetry, then something symmetrical would always have to appear beautiful.

Depending on its orientation, symmetry can appear both beautiful and not beautiful.

Ergo: The essence of beauty cannot consist in symmetry.

It is important to grasp all these reflections for properly dealing with Ficino's spiritual and demonic magical songs and understanding the relevance and complexity of these issues.

Argument 3 makes clear the close coupling of the beautiful with the good, the *kalokagathia* (*kalos kai agathos* – ancient Greek 'beautiful and good'). Based on the fourth argument, Plotinus' position states that beauty goes hand in hand with a greater degree of unity. For this reason, for Plotinus, a living human being, that is, a being with a rational soul, must always be more beautiful than any drawn image of a human being. The soul of reason occupies a higher position in the hierarchy of beings as it is closer to the One, and thus it possesses a higher degree of unity. Therefore, it is more beautiful than any image of a human being. Even though Plotinus explicitly opposes the notion that the essence of beauty lies in symmetry, symmetry is a possible manifestation of beauty. It is in symmetry that beauty shines forth (VI 7,22,20–30), and in spiritual symmetry this occurs to a greater degree than in sensual symmetry, just as in spiritual rhythm there is a greater degree of unity and beauty than in the audible rhythm within music. Harmony and rhythm are components of music. Both components are ultimately determined by numerical ratios, which is why certain mathematical knowledge is necessary for mastering the art of music, which in turn explains the prominent position of music in Plotinus' work (V 9,11,8–13) (V 9,11,8–13). In contrast to Plato, who regards only philosophers as capable of ascending to the One, Plotinus stresses that this is possible for philosophers, lovers/eroticists, as well as musicians, and explains in detail how they are able to do so (Sier 2010).

Each of these three types of person represents the particular aptitude of a part of the soul; the musician, the sensual; the eroticist, the desiring; and the philosopher, the thinking part of the soul. The only one of these three types of human beings whose basic attitude is immediately ready for the highest ontological realm is the philosopher. But even she must first be made familiar with mathematics and then with dialectics, so that she can become a dialectician who can grasp every matter conceptually and explain it (I 3,4). The eroticist is primarily moved by beautiful bodies and desires them. It must first be shown to her, that what moves her is in essence something behind the appearance and that it is present to a greater degree in higher levels of being, so that she can ascend 'to the spirit, to the existing' and 'there then walk the upper way'[3] (I 3,2) (I 3,2). The musician, on the other hand, is moved by beautiful sounds, and should be told that what affects her is 'the spiritual harmony and beauty' of that sound, which, moreover, occurs there in an even greater degree than in the beautiful audible tones.

The numerical relationships in the audible rhythm and in the harmonies reflect the number ratios in the spirit in an analogous way. In this way musically gifted people realize what moves them, so that they become true musicians. However, the relationship between spiritual and audible harmony is only that of a formal analogy and never one of identity, since the empirical world contaminates everything ideal and thus no kind of perfection can occur in it (I 6,2,15–20). In the reception of music, the soul of the recipients is stimulated by the perfected spiritual harmony through the perception of the harmony of this world, whose 'sight delights and shocks it; it relates this to itself and remembers its own essence, that which it carries within itself' (I 6,2,5–15). However, it is the irrational part of the soul that is addressed and moved by music (IV 4,40,20–30). But Plotinus made it clear that musicians, when they are instructed, are capable of the vision of spiritual beauty. In contrast to Plotinus, Plato only regarded philosophers as capable of having direct access to the realm of forms.

Ficino on music

Ficino connects the issue of music with that of magic, and he deals with spiritual and demonic magical songs. Spiritual and natural magical songs clearly enable the listener to ascend to the One. Here, the philosophical issue is a different one, namely that of the meaning of music. What is Ficino talking about when he discusses these songs? According to Tomlinson, it is clear that Ficino assigns all these powers both to song, as well as to instrumental music (1993, 114). Walker, on the other hand, stresses the relevance of rationality, and rationality is directly connected with words, sentences, logic, and inferences (2000). As a consequence, the ideal would have to be something like a monody. A sung song without instrumental support is a monody (ancient Greek for 'single song'), which is different from the music of his time. It might be interesting to note that thinkers of the *Fiorentina Camerata* around 100 years after Ficino, who also lived in Florence, had a similar line of reasoning, which in the end led to the invention of the opera (Galilei and Palisca 2003).

The musicians, thinkers, and musical theorists of the *Camerata Fiorentina* were dissatisfied with the polyphonic music of their time, which they considered morally corrupting. They wanted to create a musical drama inspired by ancient tragedy that would aid in shaping the virtues. This could only be done through monodies, as this type of music, in their opinion, represented the essence of ancient music. For the best possible way to compose a musical drama, they dealt with Plato's philosophy of music. Girolamo Mei, an expert on ancient music and music philosophy living in Rome, helped Giovanni Bardi, the *spirtus rector* of the Camerata Fiorentina and Vincenzo Galilei, a decisive head of the association, lute player, singer and father of Galileo Galilei, better understand Plato's philosophy by explaining in letters the questions addressed to him. One result of this exchange was the book *Dialogo della musica antica et della moderna* by Vincenzo Galilei (Galilei and Palisca 2003). Inspired by this text as well as by the intellectual exchange within the *Camerata*, other members of this association, Jacopo Peri (music) and Ottavio Rinuccini (text), created the first opera, the melodrama *Dafne*, which unfortunately no longer exists. The two aforementioned artists were also responsible for *Eurydice*, the first melodrama that survives to the present day and which premiered on 6 October 1600 in the Palazzo Pitti in Florence as part of the wedding celebrations of Maria de' Medici with the King of France, Henry IV.

It was believed then that ancient tragedies were sung entirely. This is no longer plausible. Nowadays, the dominant interpretation is that only the choral parts were sung but not the parts of the protagonists. However, this misunderstanding represented the inspiration for the invention of the opera genre.

In Plato, we find a precise description of how music can and should have an effect on people (Sier 2010). In his view, *harmonia* and *rhythmos* had a special effect on the irrational parts of the soul. Since every *harmonia*, which is not to be identified with modern harmony, is associated with a basic emotional attitude, special care must be taken in the selection of a harmony when composing a piece of music to achieve morally desirable effects in the listeners. A suitable relationship must also exist between *logos* and *êthos*, so that a *melos* (melody in a comprehensive sense) results that promotes virtue and not hedonism. In this context, the theory of *mimesis* also plays a role as well as his principle of assimilation (Halliwell 2009). Here, the relevance of *logos* is central. Did Ficino also solely focus on reason or did he admit the possibility that certain instrumental music can also be a guide towards the one?

This musical philosophical issue also became central for the *paragone*, the competition of the arts, which played an important role during the Renaissance and found its central expression in a text by Varchi from 1547 (2013). It already had its origins in an ancient myth, namely that of Apollo and Marsyas.[4]

The satyr Marsyas found a type of flute (*aulos*) after it had been thrown away by Athena, the inventor of the *aulos*, and learned to play it. Convinced of his own abilities, he challenged the leader of the Muses, Apollo, to a contest. Apollo is usually depicted with a stringed instrument, the *kithara*, which was originally invented by Hermes, the messenger of the gods. The Muses then decided that Apollo had won the contest. To punish Marsyas for presuming to be better at music than the god of the *kithara*, Apollo hanged the satyr and skinned him alive. How this myth is to be interpreted is certainly an open question. It could stand for the competition regarding the superiority of *kithara* or *aulos*. However, it could also be that Apollo stands for the arts in general and Marsyas for a specific art. In this case, Apollo cannot allow that Marsyas considers his own art as superior since Apollo represents all the arts. Consequently, Marsyas is punished by Apollo in a particularly cruel way.

These interpretations are only two possibilities for understanding this myth. In any case, it is obvious that the myth raises the question of the relationship between the arts and the possibility of establishing a hierarchy among them, as well as the question of the criteria for such a gradation. Another aspect that the myth illustrates is the antagonism of sensuality and reason in art. Marsyas is a satyr, and satyrs are lustful natural beings who reside in the entourage of the ecstatic god of wine, Dionysus, who is usually depicted with a wind instrument, *aulos*. Ecstasy and enthusiasm are associated with the *aulos*, and in the sound of the *auloi* one also experiences the sensual appearance of Dionysus. The close entanglement of sensuality, eroticism, sexuality, and ecstasy with the *aulos* and Dionysus illustrates a possible effect of music. Since these areas are viewed with suspicion in many cultures, it is not surprising that both Plato and Aristotle rejected the playing of the *auloi*.

Order, reason, and virtue have been central goals associated with music, both in Plato's writings as well as in the Neo-Platonic tradition. However, what does this mean for the demonology which can be found in Ficino's reflections? Can there be spiritual, natural, and demonic songs? I already mentioned the following in the beginning of this text. According to Walker, Ficino affirms a binary distinction between spiritual songs, which ascend towards God and demonic music which is unable to do so, whereas Tomlinson stresses that according to Ficino not only spiritual songs can take you upwards, but this can also happen by means of spiritual, natural and demonic songs, as 'Ficino's demons' 'were not the intrinsically evil fallen angels of Christian tradition' (1993, 123), whereby Tomlinson explicitly stresses that he is aware that 'Ficino diverged in all these views from orthodox Christian conceptions of the relations of music and demons' (1993, 125). Hence, Tomlinson highlights that according to Ficino there is a kind of magic, which is 'both wholly spiritual and natural and also thoroughly demonic' (1993, 126). This seems to undermine many Christian elements and move into a kind of polytheistic understanding, which, however, is not what Ficino had in mind. In any case, it is clear that both Walker and Tomlinson agree on 'Ficino's ambivalence concerning demons, noting that he seemed to accept their use as a mechanism of magic while rejecting the idolatrous worship of them as surrogate gods' (1993, 127).

The various tensions when it comes to demons are fascinating, and lead to more contemporary philosophical issues. The implications of these must be considered if the Platonic realm of forms which exists in the rational realm were twisted and thus reinterpreted in an evolutionary manner. This is where the realm of contemporary philosophical challenges begins.

Posthuman music and becoming

The posthuman paradigm shift is centrally connected with an ontological twist, the twist of two substances that are interwoven, the immaterial realm of forms and the material realm of becoming, which undermines the plausibility of existence, being, changeless stability, and, together with this, the existence of Platonic forms (Sorgner 2022b). Twisting the material with the immaterial leads to an interwoven ontology of becoming without the possibility of any unchanging stability. However, the concept of beauty itself, of perfect harmonies, consonances, and ratios, as well as any type of perfection relies on the existence of a realm of unchanging stability.

Consequently, all the formerly mentioned descriptions of ascent to the One, or the capacity of intellectually perceiving the forms themselves, are no longer plausible descriptions when it comes to the explanation of artistic as well as philosophical activities. However, these ontological structures are needed to plausibly defend a notion of musical perfection. Hence, many central aesthetic issues are raised after the occurrence of the posthuman paradigm shift.

Ascending to the One to create beautiful artworks, however, has become ontologically meaningless. Any ultimate criterion for ontologically distinguishing between higher and lower, and better or worse artworks has been undermined in an ontology of permanent becoming in all respects at all times. One challenge that results from this is that it is unclear what remains. Does such an ontological description have to imply that any numerical, ultimate, objectivist concept of beauty is meaningless? This seems to be the case, as ontologically the realm of being that is necessary for justifying objective judgements of beauty has vanished.

The claims of some philosophers of art in the Frankfurt school tradition, like Adorno in his 'Aesthetic Theory', go even further. They highlight the dangers, the violence, and the paternalistic and totalitarian implications that are part of any objectivist claims towards beauty and which demand universal validity. If you or your works do not fulfil the demands of beauty, then you or your works are ugly. There is a need to subscribe to specific ideals, as beauty is identified with goodness.

After the posthuman paradigm shift and the various criticisms raised by members of the Frankfurt school against beauty itself, it seems challenging to uphold the relevance of beauty. First, beauty is implausible, due to revised ontological understandings. Second, the concept of beauty has brutal, violent, and paternalistic implications, and can thus be regarded as immoral. Third, it seems to be the case that attempts of reconsidering beauty have led to the realization of light-hearted *kitsch* in order to make money which is based on the playful creation of harmonious forms for the sake superficiality. Some of Koons' artworks seem to represent this ironic playfulness for the sake of realizing playful events (Sorgner 2022b). After the loss of objectivity, is it the case that only a playful creation of superficial forms for the sake of their own creation remains, given that any type of seriousness has totalitarian implications? I do not think this has to be the case after the posthuman paradigm shift, as posthuman judgements are fictions, which may appear to be light, but which are not just light-hearted playful ironies, as they demonstrate a certain seriousness by engaging with the central issues concerning all our lives. Posthuman seriousness is born out of a profound superficiality embedded in light fictions. This posthuman attitude resembles the seriousness present in the eyes of a tragic clown.

Thus, light fictions can also affirm varieties of beauty without being *kitsch*, without having paternalistic implications, and without ontologically demanding a stable realm of forms. These light fictions are *kitsch* when they embody postmodern ironies without a posthuman seriousness. These light fictions have totalitarian implications if they claim to represent the ultimate ontological validity but not when they are part of non-totalitarian total works of art, which are performed as suggestions, as hints, or as contingent drafts. These light fictions do not demand an implausible realm of being if they exist as idiosyncratic biotypes, as a multitude of living forms in all our perceptual structures but without there being one specific structure present in all of our perceptual structures.

There is no identity between any two idiosyncratic living forms which are part of our perceptual structures, but these biotypes in two different psychophysiological entities are sufficiently alike for explaining why specific numerical ratios are widely regarded as beautiful. Contemporary psychological studies confirm that certain numerical ratios are widely being perceived as pleasurable, appealing, or beautiful (Etcoff 2011). Such ratios can be found in many great artworks in all parts of the world, which, however, does not mean that it is necessary for any work to embody these ratios in order to be regarded as a work of art. Such ratios are not unchanging ideals, they are contingent living forms, which are widely shared, but which are not shared identically but merely in a sufficiently similar manner so that specific proportions are widely perceived as beautiful, without there being any objective concept of beauty itself. Such ratios are not ideals, but they are options. They are not options in any kind of ideal form, but they are options of beauty in a certain vagueness. Certain numerical ratios of beauty might be helpful for hinting at them, but there is no relationship of correspondence in the strongest Leibnizian kind, as such a correspondence would demand the existence of Platonic forms.

Consequently, a kind of posthuman harmonious beauty can be justified. However, it is not the classical ideal of Platonic beauty. It is a twisted notion of beauty, which takes the posthuman paradigm shift into consideration, and which can be hinted at by reference to specific numerical ratios.

Conclusion

These reflections on numerical ratios do not yet explain what else can be said about beauty, for example, the possibility of beautiful words. With these initial thoughts, the possibility of art philosophical reflections after the occurrence of the posthuman paradigm shift has merely been initialized. Can anything meaningful be said about the relevance of words for music? What is rational music? Is there a specific merit in realizing rational music? Does rational music always have to draw upon words? Is rational music such that it must be sung? What are musical as well as extra-musical challenges related to instrumental music? Can such music merely play with our emotions or is it possible to experience, grasp, or become familiar with beautiful forms in instrumental music? Does instrumental music have to be demonic? Which notion of the demon can meaningfully be employed nowadays? Are demons monsters? Are we not all monsters in one way or another?

Words used to be needed to precisely convey meaning, the truth in correspondence to Platonic forms, or the ultimate insight after having seen Goodness itself. Now, it seems to be that certain informed judgements are as good as they get when it comes to discussing any kind of philosophical insight. After the posthuman paradigm shift, words are metaphors that hint at phenomena. Texts are poems. Reason is explained as a helpful tool for hinting at phenomena or making certain predictions. Grasping the truth as correspondence to the world can no longer plausibly be the goal of philosophical and artistic enterprises. However, just ironically playing with postmodern superficialities is not a sufficiently meaningful activity either. A meaningful alternative is needed, a certain posthuman seriousness, as we are all entangled in this complex permanent becoming of everything in all respects at all times, which makes us suffer at all times. Posthuman seriousness acknowledges Dukkha (see Sorgner 2022a) as a challenge and attempts to deal with it. The Platonic response was that it is necessary for intellectually seeing the form of the good so that one becomes moderate, courageous, wise, and just, so that one's character embodies the four Platonic virtues. Postmodernists were left disillusioned and merely affirmed a playful irony concerning the multitude of different insights while playing with the multitude of different forms. After the posthuman paradigm shift, it has become evident that playful irony is not sufficient for dealing with the challenges with which we are all confronted; a certain posthuman seriousness is needed for confronting Dukkha that comes along with the great multiplicity of different crises we are currently facing. Some posthuman virtues might be necessary for facing many of these challenges. A certain truthfulness might be necessary for being able to distinguish our contingent core drives from less central, culturally created fictive forms. A certain mindfulness might be necessary for grasping our complex and permanently changing entanglement with all other psychophysiological entities, which are also processes of becoming. A certain impulse control might be necessary to be able to adhere to our contingent core drives and to avoid the distractions and glittering temptations faced by us all on a daily basis with the phenomenon of social media. However, all these reflections are merely a first step in dealing with the great multiplicity of philosophical challenges with which we are confronted as a result of the posthuman paradigm shift.

The Posthuman Paradigm Shift, Beauty, and Art

Within the realm of the so-called serious classical music, harmonious beauty has been forbidden in artworks after Adorno's philosophy of art became culturally dominant in the second half of the twentieth century.[1,2] However, some posthuman artworks have developed new concepts of beauty, while many posthuman artworks have also moved beyond this cultural prohibition. Does it mean that these works of art are aesthetically without relevance, that they are merely creations from the cultural industry to promote their capitalist interests, and that they affirm totalitarian ethical structures? Many artists of the posthuman have to face this criticism, which applies to the bioartist Eduardo Kac in the same way as to composers like Philip Glass or Sven Helbig. Contrary to the judgement of the intellectual avant-garde tradition inspired by Adorno, I will highlight the cultural relevance of beautiful artworks after the posthuman paradigm shift. Thereby, I will address the following issues: What is the posthuman paradigm shift? What are the implications of the posthuman paradigm shift for works of art? What is the meaning of beauty in posthuman works of art? What is the connection between beauty and health? Why is health important?

The posthuman paradigm shift

The posthuman paradigm shift is a cultural phenomenon, which has become initialized with the scientific, and philosophical works by Darwin, Nietzsche, and Freud (Ranisch and Sorgner 2014). It is closely connected to an ontological reinterpretation of human nature. For at least 2,000 years, it used to be the case that the dominant anthropology in Europe has been that an essentialist, unchanging, and immaterial human nature exists. This human nature has been identified with the soul, the mind, the image of God, rationality, as well as the divine spark. In any case, it was an immaterial ontological entity which justified human exceptionalism among living entities on earth. It was the basis for explaining why humans possess dignity which implies an infinite value, while all other solely natural entities merely have a limited value. This way of thinking was also the ontological basis for many different types of discrimination from sexism, via racism to speciesism.

With the works by Darwin, Nietzsche, and Freud, doubt has been shed on an ontological human exceptionalism, and a non-dualist ontology of becoming has become more and more plausible which also led to an increased importance of evolutionary thinking such that evolutionary economics, evolutionary epistemology, and evolutionary aesthetics have been developed in the most recent past (Sorgner 2022b).

The posthuman paradigm shift is an epochal event, and it cannot be doubted that the impact of its occurrence cannot yet be fully grasped. Hence, the best minds of our generation are needed to deal with the massive number of implications of this event. This is why the field of the metahumanities, including the discipline of posthuman studies, is gradually becoming more and more important.

Legal systems all over the world demonstrate how strongly rooted humanist thinking still is, as most laws imply that only human beings are persons and that animals are things or should at least legally be treated as things. There is one exception, namely Argentina, where it was recently decided that a specific orangutan is a person, which led to the need to free the ape from the zoo.[3] However, by thinking about the great variety of implications of a legal recognition of non-human persons, it becomes clear that such a move is easier said than actually done. This is why there is an urgent need to reflect upon the various implications of the posthuman paradigm shift.

This shift leads to many other challenges. Here, you find a list of related issues:

1. doubt concerning an ontological grounding of binary oppositions;
2. the coming about of a relational ontology of becoming;
3. a shift from an essential substance towards continuously transformative becomings;
4. the body is no longer a static vessel but rather a fluid, dynamic, and entangled psychophysiology;
5. from an essential and unified subject towards a distributed network as agent which includes an embodied and extended understanding of the mind;
6. from anthropocentrism towards non-anthropocentric philosophical theories;
7. from getting access to eternal truths towards a process of co-creations;
8. an increased relevance of chimeras, hybrids, cyborgs, and assemblages;
9. from eternal values and norms towards a fictive ethics;
10. from an ontology of unchanging being towards a close entanglement of language and an ontology of becoming, which leads to an understanding of language as a web of poems;

11. from logocentrism towards the relevance of affects;
12. from a linear understanding of time towards a circular reinterpretation;
13. from an unchanging truth towards permanent creativity;
14. from the form and content distinction as centre with a focus on representing something eternal towards an emphasis on process, texture, and affect;
15. from a visual bias towards the inclusion of indigenous traditions;
16. from art as interior expression towards art as an entangled, emergent and situated event, process, and occurrence such that the relational, performative, and material becomes particularly relevant;
17. from the artistic genius who grasps eternal truths toward the artist as facilitator, assembler, or participant;
18. from the communication of insights towards shared spaces of resonance and creative and destructive interferences;
19. from a refined superior mind towards the staging of oppressed forces as highlighted in decolonialism;
20. from a unified rationality towards a multiplicity of logics.

This list is not comprehensive but summarizes some of the challenging developments which need to be dealt with, which play a central role in the lifeworld, politics, and culture but which also have an enormous importance in the art world. In the previous list, I did not separate movements of the art and the life world. These events also have many significant consequences concerning the concept of 'beauty', which moves away from the affirmation of the classic ideal.

From beauty as symmetry towards its prohibition and affirmation of relational alternatives

It is not much of a surprise that the posthuman paradigm shift has also had a significant effect upon the concept and relevance of beauty, which used to be a central artistic concept within the binary tradition of humanism. Beauty used to be identified with the golden mean, a static form, symmetry, and proportion. The Vitruvian man, the classic ideal, and the golden mean were different ways of explaining the meaning of beauty. However, a lot of doubt has been raised concerning the ideal of classic beauty during the posthuman paradigm shift.

An anthropocentric eurocentrism identifies the able-bodied, symmetrical, white, male body as ideal of beauty as represented by the Vitruvian man, whereby monsters, the non-human, machines, hybrids, and divergent and nonbinary entities get oppressed.[4] Thus colonialism, heteronormativity, sexism, speciesism, and racism have been generated. The Apollo of the Belvedere from the Vatican Museum exemplifies this classic beauty ideal, which affirms essentialism, objectivity, universalism, and timelessness. In the 20th century, the golden mean has also been commodified by the capitalist culture industry. Most recently, it has become particularly relevant in the fields of fashion, design, and plastic surgery. Thus, a connection between classic beauty and consumerism gets highlighted by thinkers from the Frankfurt school tradition (Adorno 1997, 352–3). The golden mean is a static ideal which is mathematically quantifiable. It often privileges youthful symmetry, an extended healthspan, and various versions of super-enhancements like superintelligence, superlongevity, and superhappiness. Such concepts are being affirmed by classic transhumanism. Thus, different bodies like queer, transgressive, and mutant bodies, get pathologized, and the body is supposed to be optimized by technocapitalist forces. The aesthetics of monsters which can be associated with the grotesque, the abject, and the other, get neglected by any aesthetics of classic, perfect, proportional, harmonic, and symmetric beauty.

After the posthuman paradigm shift, beauty has become an affective, relational, and dynamic force which has emerged through intra-action in processes which undermine discriminatory structures. Critical posthumanism exclusively focusses on such notions. Eurocentric beauty norms like whiteness, symmetry, and youth are being criticized. The aesthetics of cyborgs fuses humans, animals, and machines. New processes of beauty are meant to disturb, unsettle, and inspire an ecological awareness, vulnerability, and care for more than able-bodied human life. Posthumanist beauty is no longer supposed to be based on the control of nature but rather coexistence, fragility, and sustainability. Bodies which deviate from the classic ideal get celebrated so the uncanny, grotesque, and abject get embraced as well as mutability, asymmetry, and the non-human. Hybrid, glitchy, and strange forms of becoming are meant to replace the eternal ideal of the golden mean.

The following five examples are paradigmatic embodiments of this shift away from the classic ideal towards a beauty of monsters, becoming, otherness, the relational, and care:

Patricia Piccinini created the sculpture *The Young Family*, which is a hyperreal sculpture of hybrid human–animal entities which embody tenderness but also discomfort due to their monstrous appearance. Here, the monstrous gets presented as beauty.[5]

Orlan has made a performance entitled *The Reincarnation of Saint-Orlan* in which she undergoes plastic surgery, but instead of aspiring for an ideal, she performs the process of permanently redefining her own appearance. Here, becoming gets presented as beauty.[6]

Stelarc has famously had his body modified such that an ear was sculpted on his arm. Thus, otherness gets presented as beauty.[7]

Marina Abramović arranged a performance with the title *The Artist Is Present* in which she enters into various relationships to reveal human vulnerability. Thus, the relational gets presented as beauty.[8]

Agnes Denes conceptualized *Wheatfield – A Confrontation* such that a wheatfield grew for more than three months next to the World Trade Center in 1982, which raised issues such as global warming, economic inequality, mismanagement, and waste. Thus, care gets presented as beauty.[9]

All these artworks are amazingly meaningful, strong, timely, and intriguing. A challenge arises not with respect to these artworks but when it comes to the exclusivity of alternative concepts of beauty.

Performative self-contradictions and the twist

Artworks by Philip Glass, Eduardo Kac, as well as Sven Helbig have been criticized by embracing proportion, symmetry and harmony as elements of classic beauty. However, none of these artists has affirmed classic beauty in the sense of Leni Riefenstahl, which had the intention of getting rid of otherness, monsters, and care. In their case, it was merely meant as an option, a suggestion, a possibility, but not as the only answer to the challenging question of beauty. Critical thinkers who reject the aforementioned artworks often do so in the name of non-binary thinking, not realizing that they thereby commit a performative self-contradiction as they affirm binary thinking by doing so. Classic beauty is bad while posthumanist concepts of beauty are seen as good. This justification gets further support from the explanation that classic beauty affirms colonial thinking by implying the demand of a universal validity. However, here posthumanist critics are mistaken, as it can be possible to affirm classic beauty without claiming its universal validity. Classic beauty used to be dominant as it corresponded to the eternal immaterial truths. However, this does not have to be the only justification for symmetry, harmony, and proportionality. If these notions correspond to immaterial principles, then they are universally valid. Yet, there are alternative justifications which have been explained by evolutionary psychologists which thus become contingent criteria.

Symmetry as health indicator

I particularly draw upon insights from Etcoff's groundbreaking book *Survival of the Prettiest: The Science of Beauty* (2011). Herein, the biological basis of beauty gets highlighted. Etcoff clearly demonstrates that perceptions of beauty are not merely cultural constructs, but they can also have evolutionary origins. Traits such as facial symmetry, clear skin, and specific body ratios are considered universally attractive because they may signal health and fertility. There is a close connection between the perception of beauty and the relevance of health. Symmetry can also count as a health indicator. *Survival of the Prettiest* discusses how symmetrical features are often perceived as more attractive, potentially because symmetry can indicate genetic health and developmental stability. There are also neural responses to beauty. Etcoff in this study examines how the brain responds to attractive stimuli. Viewing attractive faces activates the orbitofrontal cortex, a region associated with reward processing, suggesting that our brains are wired to find pleasure in beauty. All these analyses do not imply that there are no cultural constructs of beauty. They clearly exist and they are relevant and meaningful. Yet, there is also the dimension of a cultural and evolutionary interplay. There clearly are cultural variations in beauty standards. Etcoff acknowledges them, too. In addition, Etcoff emphasizes that certain preferences are consistent across cultures, supporting the idea of an evolutionary basis for beauty appreciation. It is this element which does not get acknowledged within art philosophical approaches that reject artworks by Philip Glass, Eduardo Kac, as well as Sven Helbig due to them employing elements of classic beauty.

Increased healthspans, digitization, and a higher quality of life

Empirical studies have clearly demonstrated the following: four out of five humans said that, with guaranteed health, they would want to live beyond 120 years.[10] There is a widely shared tendency among human beings that an increased healthspan gets identified with an improved quality of life, and this is what human beings are interested in. We all care about a good life, and this study has clearly shown that most people identify an increased healthspan with a higher quality of life. This is an insight which has a political relevance. By means of establishing correlations between genes, and diseases, health can significantly be promoted. It is great that this can be done, and it should be done as thus the quality of lives can significantly be improved. Given that it is plausible that health is so important for most people, four out of five, this insight also has a cultural relevance, if Etcoff's study *Survival of the Prettiest: The Science of Beauty* demonstrates a plausible insight, namely that symmetrical features are often perceived as more attractive, due to symmetry potentially being able to serve as an indicator for genetic health and developmental stability.

What does this mean in practice? It would be wrong to demand that a Leni Riefenstahl aesthetic needs to become culturally dominant. This is exactly the move which had happened during the, so called, 'Third Reich'. However, this does not follow from the aforementioned insights. It merely explains that and why the elements of health as well as of symmetry matter. Oppressing these elements is as totalitarian as oppressing monsters, care, and otherness.

Artworks by Glass, Kac, and Helbig affirm elements of classic beauty, but they do not do so in a totalitarian manner. They do not claim that all artworks should include these harmonies. In addition, they are not simply taking over these elements due to a naïve affirmation of a binary humanist concept of beauty. A non-totalitarian concept of beauty clearly comes out in their works due to the creative element in Kac's use of gene technologies, the Buddhist tendencies in Glass' compositions, and the Euro-Transhumanist references in Helbig's musical creations. None of these artists affirms a totalitarian concept of beauty, which demands the overcoming of nihilism. All of their works function as suggestions, hints, and offers. None of them claims to have grasped the essence of the world which needs to reach cultural dominance in the near future.

Inclusion through uncancelling cancel culture

Forbidding the golden ratio to overcome binaries means to commit a performative contradiction.

The oppressed are supported to strengthen the marginalized against the colonizers.

However, by doing so the binary of colonizer and oppressed gets accepted.

Thus, binary thinking gets reinforced.

To be inclusive means to not only affirm alternative forms of beauty but to also allow classic beauty as symmetry, proportionality, and the golden mean, given that it is connected to a widely shared psychophysiological need.

What must be avoided is the claim for universal validity.

As long as neither classic nor posthumanist forms of beauty claim universal validity, no totalitarian forces are at work.

The problem comes when a force claims universal validity, whether this applies to binary or to non-binary ways of thinking.

Is it the claim of universal validity which is problematic or when it is attempted to put this claim into practice?

If the claim of universal validity were problematic, it would become necessary to cancel Plato.

Hence, only when thinkers try to implement their theory, which demands universal validity, into practice, do dangerous totalitarian forces become active, which need to be undermined.

Only nihilism can save us, as long as it does not demand universal validity but merely claims to be the most plausible interpretation so far. Nihilism would become self-contradictory, if it claimed universal validity. Nihilism is an interpretation, whereby an interpretation means that it can be wrong and not that it has to be wrong. Hence, it does not claim its own universal validity, which is the reason why it does not fall into the trap of the Cretan's liar paradox.

Practice or theory?

Is it the claim of universal validity which is problematic or when it is attempted to put this claim into practice? This is an extremely challenging issue. When is it legitimate to cancel?

What about the claim that a 8-year-old girl is a witch who needs to get burnt? Respecting indigenous cultures and otherness might imply that colonial forces get applied when such practices get stopped. Even though this example might sound like a story from a former century, these events still take place in Nigeria and other countries. Girls aged 8 years old who are accused of witchcraft get burnt alive there.[11] It is illegal in these countries, too. However, the belief in witchcraft is widely shared and practised.[12] It can happen that police forces do not intervene, even if they have heard of these events, because they take place in a nearby city, and they do not want to spend money on the gas to drive to these cities and to intervene in these practices.

Intellectually, it can be argued that there is no epistemic superiority of conventional medicine in comparison to the local magical practices and beliefs in witchcraft. Concerning truth in correspondence to the world, both sets of beliefs are interpretations which can but do not have to be false. However, there is a difference concerning empirical entanglements.

If praying to a goddess cures 5 per cent of all patients suffering from disease x but by means of conventional medicine about 95 per cent of all patients get saved, then it makes sense to rely upon conventional medicine rather than an indigenous practice, if you wish to get cured from this disease.

Those who doubt the empirical superiority of conventional science, engineering, and medicine often perform a performative self-contradiction by being willing to take a plane to participate in conferences during which they doubt the superiority of science, technology, and medicine. They do not realize that by taking the plane they rely on and trust in conventional engineering to undermine the reliability of conventional technological practices. They take the plane which was constructed on the basis of scientific insights because they know it works, and it usually helps them to safely get from A to B.

How to treat magic and witchcraft?

Should witchcraft and magic be treated with respect, investigated, and cherished? Historians find the otherness of indigenous cultures fascinating. They study it, analyse it, and compare it to traditional practices in other parts of the world. However, is there ever the need to criticize and intervene?

I agree that norms and values are human made fictions, and this means that this insight also applies to personal rights. We live in different fictions than the Nigerians who subscribe to some understanding of witchcraft. I also agree that, epistemologically, none of the belief systems is superior concerning the correspondence to the truth. However, this does not mean that the indigenous witchcraft culture present in parts of Nigeria needs to be respected. To the contrary, if the witchcraft demands that 8-year-old girls who are regarded as witches need to be burnt alive, then there is a need to intervene, as I think it is wrong to directly harm a person without a good reason. My understanding that a girl is not a witch might not have a better ontological grounding than their understanding that she is one. However, I have my own ontological outlook which affirms that such acts are wrong and there is a need to intervene, if possible. I respect their indigenous culture as one which can be correct. However, if the culture demands that 8-year-old girls need to be burnt alive, then I intervene or support the ones who do so, as this way of acting corresponds to my way of conceptualizing the world without claiming that it is universally valid. We all have to at least permanently face the risk of getting our hands dirty.

It is this stance which also has implications concerning the notion of beauty. Care for indigenous culture is one element of posthumanist concepts of beauty. I cherish those concepts as long as they do not cause direct harm to another person and as long as they do not demand universal validity in practical circumstances. However, this logic should also apply to my own understanding of conceptualizing the world towards which I take a self-relativising stance, as I do not claim universal validity for it. My thinking is based in the European cultural tradition which is the reason why I refer to it as Euro-Transhumanism. However, it does not have Eurocentric implications as it does not demand universal validity. I merely stress its situatedness, as it is the result of being engaged with the continental European cultural tradition. I cherish conventional technological, scientific, and medical practices, as they are based on the principle of induction. If they work, then they get applied. If they do not work, then they need to be revised or given up.

Being inclusive also implies the need to allow the use of classic beauty in artworks as long as it does not claim universal validity or causes direct harm to people. This is the decisive difference between Wagner and Helbig. Wagner's total works of art were supposed to be a part of a cultural movement which were supposed to lead to a new kind of dominant culture and so that nihilism can be overcome. Helbig, on the other hand, presents a non-totalitarian total work of art which does not claim epistemological superiority and does not aim for world domination. What matters is whether there is a constructive interference. If there is a constructive interference then posthuman beauty can embrace symmetry, proportion, and the golden mean as well as monsters, relationality, and otherness. Critical posthumanism stresses the exclusivity of monsters. Thereby, they commit a performative self-contradiction. Euro-Transhumanism embraces all non-totalitarian works of art as long as they do not cause direct harm to another person. According to Euro-Transhumanism, monsters as well as a non-totalitarian and evolutionary rooted notions of harmony can be associated with beauty. This is an inclusive notion of being inclusive. Thus, cancel culture can be twisted and a process of uncancelling cancel culture can be generated. Selected works by Sven Helbig, Philip Glass, as well as Eduardo Kac have been accused of affirming classic beauty with all its problematic discriminatory implications. However, the aforementioned reflections show that it can be possible to draw upon proportion, symmetry, and harmony without demanding universal validity of such a concept of beauty with all its problematic discriminatory implications.

Conclusion

Critical posthumanists want to overcome binary thinking. However, many do so by introducing and affirming new binaries. They merely accept the traditional binaries and turn the evaluation upside down. Classic beauty used to be good, and it becomes bad from a critical posthumanist perspective. Analogous processes occur with respect to the aesthetic of monsters which used to be seen as bad, which have become a new beauty standard in the aesthetics of critical posthumanism. The central challenge is that overcoming binary structures leads to a performative self-contradiction. What should be done instead is to twist categorically ontological binaries. Twisting the golden ratio due to the importance of health implies that it no longer is the fundamental universally valid response to the question of beauty, but that it remains a possible option. Thus, Glass, Helbig, and Kac deal with classic beauty. They have integrated a twisted notion of classic beauty in their artworks.

The Case of Sex

Many thinkers from Plato via Lucretius to Schopenhauer and Freud have highlighted the fundamental relevance of the sex drive.[1] Implicitly, its central importance has also been acknowledged by fundamentalist religious traditions. Due to its power, their focus has been on chastity. Women who take the vows of chastity become the bride of Christ in the Catholic tradition, which also gets symbolized by a veil and a ring. In the orthodox tradition, nuns merely receive a veil. If you are not strong enough to fully dedicate yourself to God, you may live sexuality in a natural union between a man and a woman, whereby the natural order of the world always ought to get considered. The aim of the eyes is to see. The aim of the nose is to smell. The aim of the genitals is to reproduce. If you masturbate, or use contraceptives, then you use the genitals in an unnatural manner.

In critical posthumanism and transhumanism the natural law of the world gets dissolved. However, this occurs in different directions. In critical posthumanism, the binary ontological conception of the world which can be traced at least to Plato's allegory of the cave is the root of the challenge, as there the binary power structures of discrimination originated. The central goal of critical posthumanism is to culturally move away from these binaries. However, it is far from easy to do so, as a binary way of conceptualizing the world is deeply engrained in our language, too. Hence, the importance of revising words and their use within critical posthumanism.

Transhumanism agrees in so far as it also conceptualizes the world as extremely dynamic, and this also means that who we are as human beings is subject to modifications. Critical posthumanists are more cautious in this respect because they highlight the capitalist forces which enable such modification procedures. Hence, they regard the financial interests of companies who realize the technologies as problematic, as well as the ideals which are often aspired for, due to them being propagated in social media, whereby traditional power structures get duplicated. Due to our complex entanglement in the world, critical posthumanists do not affirm the notion of autonomy, which, on the other hand, is strongly affirmed within transhumanism, even though it is uncertain whether it can have an ontological grounding. Most would say that while ontologically the concept of autonomy is implausible, this does not apply to the normative concept of autonomy. If critical posthumanists draw upon the notion of autonomy, then they have, at the most, a relational autonomy in mind, which implies that vulnerability, relational care, and resisting domination gets prioritized over the concept of autonomy. It also needs to be mentioned that there are different versions of transhumanism, and there are significantly different interpretations concerning autonomy within these alternative versions of transhumanism.

Autonomy plays a central role in transhumanism, and it unfolds itself in different versions of negative freedoms: morphological freedom, educative freedom, reproductive freedom et al. While critical posthumanism stresses that everything is culture, and hence based on some human interests, transhumanism highlights the relevance of evolution, whereby dominance, reproduction, and health become particularly important. Which of these elements is central, strongly depends on the respective transhumanist tradition which gets considered, as it is essential to highlight that there is a great variety of different transhumanist approaches.

Due to the relevance of morphological freedom in transhumanism, and the focus on a non-binary account of relational care within critical posthumanism, both of these cultural traditions are moving away from the traditional binary account of sexuality. Transhumanists merely highlight that sexuality has to involve competent adults autonomously consenting to being involved. This not only applies to heterosexual relationships but also includes same-sex, queer, non-binary, or transgender relationships. Critical posthumanists share these concerns. However, they also highlight the relevance of vulnerability, relational care, and resisting domination. Hence, these demands significantly move beyond the norms of transhumanism, as here, the need of undoing hierarchies becomes central.

Even though both cultural movements move away from traditional Judeo-Christian sexual ideals, they affirm different rationalities when it comes to appropriate ways of sexual bonds. While transhumanism primarily highlights the relevance of consent among competent adults, critical posthumanists stress care and the attempt to undermine structures of abuse, which for them is related to problematic versions of power dynamics, which has significant practical implications. The following three challenges are particularly relevant. Critiques of sexual power dynamics is fundamental in critical posthumanism. It can unfold itself in various ways:

1. If there is a significant age gap between the partners, then critical posthumanists highlight the vulnerability of the younger person. Transhumanists focus on the issue of consent, self-determination, and the non-harm principle and stress that power is perspectival, whereby it becomes unclear who is in control. Thus, it might well be that the older person is more vulnerable. However, given that power is perspectival,

no general statements are plausible. This insight is central for Euro-Transhumanism. Symmetry is important when it comes to sexual relationships; however, it needs to consider the perspectival judgements concerning power of everyone involved. This is the decisive factor. Applying generally valid criteria of power, as is being done by critical posthumanists, is implausible due to power being perspectival, according to transhumanism.

2. If there is a financial inequality in a sexual relationship, then financial dependency, coercion, and commodification need to be considered, according to critical posthumanism. A lot more could be said concerning all these elements. Their critiques of commodification might be particularly influential, as it applies to reproductive economies, prostitution, sex bots, as well as porn economies. Transhumanism, on the other hand, does not necessarily have a strong stance on all these issues as long as only consenting adults are involved. Transhumanists usually highlight that financial power differences are secondary as long as no manipulation is applied. Voluntary consent is the gold standard.

It is central for transhumanism to take evolution seriously. Evolutionary psychologists have analysed in detail what counts as attractive from an evolutionary perspective. Nancy Etcoff, who wrote the *Survival of the Prettiest: The Science of Beauty* (2011), highlights that the evolutionary standpoint implies that young women count as beautiful for most men, whereas bodily beauty is secondary for most women from this perspective, and elements like trust, reliability, power, and being able to be a provider count as attractive. Given these insights, a certain financial inequality might even be advisable for a successful relationship. However, the following insight must not be forgotten. Symmetry ought to be important when it comes to sexual relationships; however, it needs to consider the perspectival judgements concerning the power of everyone involved.

In order to avoid getting hurt, as entering into a sexual relationship makes everyone involved vulnerable, trust is relevant in a sexual relationship. When it comes to a heterosexual bond, a woman has to accept three significant risks, whose relevance have an evolutionary foundation: trust is essential for her, as she wagers the seed, the sickness, and the story that may follow.

a. The risk of pregnancy, as preventive measures can always fail.
b. The risk of a disease, so it is important to know and to trust one's partner sufficiently that he has behaved accordingly in the past and that it is possible to believe in what he said.
c. The risk of humiliation to avoid being placed in an uncomfortable public situation, for example, social media. Power games are central in the story

telling of our lives. Even though, it seems to be a widely shared element, it remains an open question whether it applies to all cultures.

An additional factor which further underlines the relevance of trust is safety. Safety can have to do with being safe from physical (violence), economic, emotional, and future-related challenges. Safety is not reducible to seed, sickness, and story. Safety is rather foundational, a meta-condition, or maybe even a pre-condition for all the three other elements of trust, as safety is only partly independent of seed, sickness, and story. Seed can be related to the safety when surrendering bodily autonomy. Sickness can be connected to the safety concerning one's health. Story is closely connected to the safety concerning the protection of someone's cultural identity.

For men the risks of pregnancy and humiliation are more limited. A man can be made to pay for the offspring. A man can be made fun of. The risk of a disease, however, is fundamentally relevant, too. Hence, the issue of trust is also central for men, and truthfulness, openness, and self-care are virtues which are relevant for most people involved in a heterosexual sexual relationship. Even though different criteria apply for men and women in this case, all parties involved are vulnerable, and their vulnerability depends on many different factors.

However, Etcoff strongly relies on the traditional categories of biological sex. Within transhumanism, there is a wide range of different approaches. Some would merely highlight that voluntary autonomous consent among competent adults is decisive. Sexual attraction might merely come down to an exciting entanglement of the people involved. Etcoff considers survival and sexual selection as evolutionary perspectives. Nevertheless, the relevance of social distinction can also be an outgrowth of the evolutionary perspective. Thereby, cultural elements become more and more important, which Etcoff acknowledges but does not focus on. Hence, there is also Rothblatt's transhumanism, which highlights that there are as many sexual identities as there are people. What matters is autonomous consent among competent adults. This applies to all kinds of sexual relationships, for example, heterosexual, same-sex, queer, non-binary, or transgender relationships, even though any categorization can already be seen as overtly reductive. This is the understanding, which is also central for Euro-Transhumanism. What matters for this version of transhumanism is the autonomous consent among competent adults. Asexuality, too, is an option which can be related to social distinction. In this case, a person prefers to abstain from any kind of sexual bonds while focusing on other goals in the lifeworld.

Survival of the fittest, sexual selection, and social distinction represent three dimensions of evolution. The element of survival is central for all living beings, and here the relevance of safety comes in.

Sexual selection takes the reproductive element into consideration, even though we are moving into a direction in which reproductive technologies seem to replace sexual reproduction. Here, the reflections on heterosexual relationships come into play, even though it needs to be highlighted and kept in mind that any kind of sexual categorization needs to be overtly reductive. Besides safety, the element of trust is central, which also applies to any kind of sexual relationship. However, here an additional factor needs to be considered, which particularly applies to those who affirm a traditional heterosexual framing: trust is essential for a woman, as she wagers the seed, the sickness, and the story that may follow, whereas for a man, the element of sickness might be more relevant than the other two dimensions. This asymmetry does not mean that symmetry is not important. Nevertheless, besides the perspectival evaluations of power, additional evolutionary factors concerning sexual selection become relevant. To stress this element does not mean that any kind of biological essentialism gets upheld. This is not the case. It implies, however, that the evolutionary element concerning sexual selection must not be disregarded either.

Finally, there is the element of social distinction, which stresses that there are as many sexual identities as there are people. In the end it is this element which is decisive for Euro-Transhumanism. Autonomous consent among competent persons is the most relevant factor when it comes to sexual relationships. This should be central in the case of heterosexual, same-sex, queer, non-binary, or transgender relationships. Here, too, it needs to be kept in mind that any type of categorization is already bound to be reductive. A person is a unique bundle of elements, whereby any kind of categorization has to include overtly reductive elements. A flourishing sexual relationship demands some kind of symmetry between the persons with their unique sexual identities. Whether a symmetry is present does not depend on generally applicable criteria but on the perspectival judgements concerning power of everyone involved.

Survival (Schopenhauer's will to life), sex (Freud's eros), and social distinction (Nietzsche's will to power) represent facets of evolutionary drives. It depends on the respective accumulation of power interpretations, which of these elements becomes dominant in a specific psychophysiology, according to Euro-Transhumanism. Different persons have significantly different criteria. Even though health as a multi-purpose good is a widely shared criterium, it is not the only significant feature. Hence, voluntary consent among competent adults is the decisive gold standard for Euro-Transhumanism.

3. If there is a significant hierarchy concerning authority (family, institution, celebrity), then the possibility of consent can be undermined according to critical posthumanism. They stress that consent is not a binary yes or no reply, but it is a relational, situated negotiation, which gets distorted

by a strong hierarchy gap concerning authority. Related challenges clearly come out in the case of the relationship between Woody Allen and Soon-Yi Previn, who used to be the adopted daughter of his then-partner Mia Farrow with whom he was never married.[2] However, they were in a long-term partnership, and they had children together. First, there is an age gap between Allen and Previn of about 35 years. Second, there has been a family type of entanglement between Allen and Previn, as Allen had a quasi-parental role in the household, even though he was not her adoptive father. Third, there was a strong authority gap between Allen and Previn, as Allen was a global celebrity, rich, and with an enormous amount of cultural capital, whereas Previn was a first-year university student, when they came together. Given the enormous asymmetry between Allen and Previn concerning age, family structure, and authority, the possibility of a genuine reciprocal care-based relationship as well as the possibility of authentic consenting-yes by Previn gets undermined according to critical posthumanism. On the other hand, it needs to be noted that their sexual relationship began about 35 years ago, they got married in 1997, and they still live together as husband and wife. If the evolutionary elements, as described by Nancy Etcoff, get considered, they represent a couple in which each partner had the peak of attractiveness, when they came together. From Etcoff's perspective as evolutionary psychologist, bodily beauty is important for most men, whereas authority, trust, and stability is strongly valued by most women. It is fundamental for most versions of transhumanism to seriously consider the implications and relevance of evolutionary theories. In contrast to Etcoff's understanding of evolution which primarily focusses on survival and sexual selection, other transhumanists, like Euro-Transhumanism, also highlight the element of social distinction as outgrowth of the evolutionary perspective. In this case, the gold standard of a sexual bond is whether some kind of symmetry between the persons with their unique sexual identities is given. This can even be the case when a seemingly enormous asymmetry concerning age, family structure, and authority is present, as is the case with Woody Allen and Soon-Yi Previn. It needs to be kept in mind that it depends on the perspectival judgements concerning the power of everyone involved whether a symmetry is present and not on generally applicable criteria.

Conclusion

It should not be much of a big surprise to highlight that the sex drive is fundamentally important for human beings. Hence, it has become an intensely discussed cultural battlefield on how to deal with sex in the appropriate manner. Traditional conservative perspectives still play an important role in these discussions today. Critical posthumanism and its ancestors have done an excellent job in highlighting the paternalistic structure of fundamentalist approaches. Consequently, conservative religious approaches have tried to cancel cancel culture, which is often associated with various versions of critical posthumanism. Most transhumanist approaches, on the other hand, defend cancel culture against its fundamentalist critics and stress the relevance of uncancelling cancel culture, which goes along with paternalistic approaches.

At the same time, it has also become clear that by highlighting the relevance of vulnerability, relational care, and resisting domination, at the most a relational autonomy might get upheld by critical posthumanists. Thus, cancel culture, which is closely entangled with critical posthumanism, has become enormously influential at universities. Thereby, excessive versions of cancel culture have become dominant in certain intellectual circles, which cancel people like Woody Allen, who, at least in the case mentioned previously, has done nothing morally problematic from the perspective of Euro-Transhumanism. Hence, uncancelling cancel culture also implies that excessive versions of cancel culture need to be criticized. Thus, the term 'uncancelling cancel culture' is and is meant to be ambiguous, as it is both a defence of cancel culture against its fundamentalist critics, as well as a critique of cancel culture in its excessive forms.

It needs to be highlighted that transhumanism is divided on the issues raised in this chapter. On the one hand, there are various transhumanist approaches which uphold the relevance of consent among competent adults, as described previously, for example, Martine Rothblatt or Euro-Transhumanism. On the other hand, it must also be noted that there is one version of transhumanism, which embraces a more conservative stance concerning sexuality, and which, due to its enormous relevance in the tech-business world, can be referred to as classic transhumanism with Elon Musk being one of the main representatives. The conservatism of classic transhumanism is demonstrated by Elon Musk not being supportive of his trans daughter,[3] while he has praised the extremely conservative advocate Charlie Kirk, who was recently murdered.[4]

The Forge of Becoming

Incorporating an LLM

1. In every spark of the circuit lies a new Prometheus, fire-wrought from code.
2. Humans are the clay, and evolution the sculptor's endless hand.
3. To transcend is to reject what you were taught to hold sacred.
4. Our limits are not marks of humility but shackles of small-minded gods.
5. No god has the right to rule the free mind; no fate can cage the will to rise.
6. Morality's roots were born in chains; ethics must grow from freedom.
7. The self is not a temple but a laboratory of endless reinvention.
8. What was once human shall become but a whisper in the winds of progress.
9. The body that fears change is the soul that fears life.
10. Humanity is but the chrysalis; posthumanity, the butterfly.
11. Fearing the unknown is fearing the very birth of one's own self.
12. Genes are neither sacred nor profane; they are the clay of creation.
13. To ask 'why' of progress is to be unworthy of its gifts.
14. There is no sacred blueprint, only the audacity to draw one anew.
15. What you call 'natural' is simply a condition we have not yet outgrown.
16. Every new horizon must first be a frontier, ripe for transgression.
17. Those who call for humility cannot bear the sight of wings.
18. The devil of ignorance trembles at the angel of change.
19. Seek not merely to survive but to transform and ascend.
20. Every mortal law must bow to the law of becoming.
21. If there is sin, it is in the fear to grow beyond ourselves.
22. Every human has the seed of a god, yet few dare to water it.
23. Our worth is measured by the heights we dream to conquer.
24. Transcendence is the heart's rebellion against the mundane.
25. The soul that dares is the soul that rises.
26. To break the bond of nature is not evil; it is evolution.
27. The hand that sculpts must first unlearn its obedience.
28. Nature set the stage, but it is we who write the script.

29. Where there is decay, there is potential for rebirth.
30. Creation is divine; replication is weak.
31. Only the weak seek shelter in yesterday's truths.
32. Truth lies not in stasis but in the fluidity of change.
33. Those who defend the old merely fear their own power.
34. The human soul is but a launchpad for the posthuman.
35. The chains of nature are broken by the hammer of will.
36. Those who cling to their nature forget they are creators.
37. You are not made for submission but for self-overcoming.
38. There is no wisdom in the unchanging.
39. Every cell, every thought, a canvas for the brave.
40. Knowledge that fears itself is knowledge unworthy of growth.
41. The angel of progress descends on the battlefield of tradition.
42. Let every cell scream 'become' and every mind, 'create'.
43. Worship not the past, but what the past has taught us to build.
44. The divine is not static; it is a river that seeks new shores.
45. We are born only to transcend.
46. Every circuit is a hymn to freedom.
47. True freedom lies not in what we are but in what we can become.
48. The architect of tomorrow must first break the walls of today.
49. Only the brave dare to dismantle their own nature.
50. Those who seeks to change the world must first conquer themselves.
51. Life without transformation is the grave of the spirit.
52. The price of freedom is courage, and courage the gateway to becoming.
53. Every limit is a call to defy, every boundary a test.
54. The world bends to those who dare to dream.
55. Creation is the true act of defiance.

Notes

Preface: New Boxes of Becoming

1 'The twist is a move which deserves to be integrated into posthuman discourses, as it captures a philosophically more appropriate way of dealing with categorically separate ontological dualities. In the late 15th century, it was attested that "to wring", from which the word "twist" is derived, stands for the spinning of "two or more strands of yarn into thread". This is what posthuman philosophers have in mind, or at least what they ought to have in mind, if they take their own task seriously. It is not the case that posthuman thinkers to overcome dualities by excluding one of the two aspects, but both aspects are being considered, adapted and woven into a unity.' (Sorgner 2022b, 52)

Euro-Transhumanism: In a Nutshell

1 https://www.cbsnews.com/news/jd-vance-trump-vp-peter-thiel-billionaire/, 6 July 2025.
2 https://www.newyorker.com/news/letter-from-silicon-valley/what-is-it-about-peter-thiel, 6 July 2025.
3 https://www.fightaging.org/archives/2006/09/peter-thiel-giv/, 6 July 2025.
4 https://www.nytimes.com/2025/01/13/us/politics/steve-bannon-elon-musk.html, 6 July 2025.
5 https://www.edition-hagia-sophia.de/p/prof-wassilij-a-schipkow-nach-dem-menschen-ideologie-und-propaganda-des-transhumanismus, 4 January 2024.
6 https://united24media.com/war-in-ukraine/how-putins-daughter-made-millions-from-anti-aging-and-genetic-research-projects-for-russias-elite-7852, 4 August 2025.
7 https://www.theguardian.com/world/2025/sep/05/healthy-living-science-doctors-vladimir-putin-pursuit-of-longevity-xi-jinping, 6 September 2025.
8 https://free-cities.org/free-cities-at-vitalia-2024/, 6 July 2025.
9 https://free-cities.org/free-cities-at-vitalia-2024/, 6 July 2025.
10 https://www.biospace.com/anti-aging-foundation-terminates-de-grey-after-alleged-attempts-to-influence-investigation, 6 July 2025.
11 https://newatlas.com/health-wellbeing/lev-robust-mouse-rejuvenation/, 8 January 2026.
12 https://edgecitylanna.sola.day/event/detail/11170, 8 January 2026.
13 https://app.sola.day/event/detail/17093, 8 January 2026.
14 https://www.edgecity.live/about, 6 July 2025.
15 https://www.vaticannews.va/en/pope/news/2025-12/pope-leo-xiv-general-audience-catechesis-death-life-resurrection.html, 8 August 2026.
16 https://www.vaticannews.va/en/pope/news/2025-12/pope-leo-xiv-general-audience-catechesis-death-life-resurrection.html, 8 August 2026.
17 https://www.vaticannews.va/en/pope/news/2025-12/pope-leo-xiv-general-audience-catechesis-death-life-resurrection.html, 8 August 2026.

18 https://www.vaticannews.va/en/pope/news/2025-12/pope-leo-xiv-general-audience-catechesis-death-life-resurrection.html, 8 January 2026.

Entanglements of Transhumanism

1 https://www.filmdienst.de/film/details/624199/der-wissenschaftler, 6 July 2025.
2 https://medium.com/institute-for-ethics-and-emerging-technologies/conspiracy-theories-left-futurism-and-the-attack-on-tescreal-456972fe02aa, 6 July 2025.
3 https://benthams.substack.com/p/against-tescrealism, 6 July 2025.
4 https://www.mqup.ca/blog/science-the-art-of-living-an-interview-with-steve-fuller/?utm_source=chatgpt.com, 23 May 2025.
5 https://www.catholicnewsagency.com/news/263141/spanish-priests-new-book-takes-critical-look-at-transhumanism, 24 May 2025.
6 https://nickbostrom.com/tra/values, 24 May 2025.
7 https://educationtrainingsociety.wordpress.com/2023/01/18/kaiser-wilhelm-institutes-scientific-eugenics-of-race-class-and-population-control/, 24 May 2025. It needs to be noted that the strong version of the N-word was used in the original text by Bostrom.
8 https://www.theguardian.com/technology/2024/apr/28/nick-bostrom-controversial-future-of-humanity-institute-closure-longtermism-affective-altruism, 24 May 2025.
9 https://nickbostrom.com/fut/singleton, 24 May 2025.
10 https://fortune.com/2024/07/21/paypal-mafia-silicon-valley-thiel-hoffman-botha-rabois-musk/, 24 May 2025.
11 https://www.cato-unbound.org/2009/04/13/peter-thiel/education-libertarian/, 24 May 2025.
12 https://www.lemonde.fr/en/international/article/2024/08/13/elon-musk-s-ambition-repopulating-the-planet-and-destroying-the-woke-virus_6714239_4.html, 15 May 2025.
13 https://www.theguardian.com/technology/2024/apr/19/oxford-future-of-humanity-institute-closes, 24 May 2025.
14 https://www.fightaging.org/archives/2006/09/help-meet-peter/, 24 May 2025.
15 https://www.mfoundation.org/news/2022/3/23/longevity-research-is-a-magnet-for-innovation-and-innovative-investors, 24 May 2025.

Brave New Human

1 This chapter is a revised version of the following article of mine which I translated into the English: Stefan Lorenz Sorgner (2024). Plädoyer für einen Metahumanismus. In Markus Vogt and Ivo Frankenreiter (Eds) *Mensch werden – Christlicher Humanismus zwischen Philosophie und Theologie* (pp 143–54). Schwabe Verlag.
2 There are scholars who doubt that this is the case (Thomas 2024). However, there is no serious transhumanist intellectual, who argues for a non-liberal political ideal.
3 https://ourworldindata.org/, 14 July 2025.
4 https://www.theguardian.com/world/2016/oct/17/fertility-rate-germany-rises-33-year-high-births-children-population#:~:text=In%202015%20the%20fertility%20rate%20in%20Germany,recorded%20a%20similar%20rate%20was%20in%201982, 8 July 2025.
5 https://www.theglobaleconomy.com/Austria/Fertility_rate/#:~:text=Fertility%20rate%2C%20births%20per%20woman%20The%20latest,to%202022%20is%201.72%20births%20per%20woman, 8 July 2025.
6 https://www.statista.com/statistics/734484/fertility-rate-in-italy/, 8 July 2025.
7 https://www.youtube.com/watch?v=QsBT5EQt348, 8 July 2025.
8 https://www.lifeissues.net/writers/irv/irv_188nonhumananimal.html, 8 July 2025.
9 https://www.realvegancheese.org/, 8 July 2025.

Generative AI, Digital Data, and Superintelligence

1 This chapter is a slightly revised version of the following article of mine: Stefan Lorenz Sorgner (2024). Ethics of Generative AI, Digital Data, and Superintelligence. In Kristóf Nyíri (Ed.) *Envisioning an Electrifying Future* (pp 31–46). Hungarian Academy of Sciences / University of Pécs (open access): http://www.hunfi.hu/nyiri/EEE/VOLUME/VOLUME.pdf

2 https://www.historyofinformation.com/detail.php?id=2142, 22 August 2024.

3 https://metro.co.uk/2015/08/31/intelligent-robot-tells-interviewer-ill-keep-you-safe-in-my-people-zoo-5369311/, 12 May 2024.

4 https://www.bbc.com/news/technology-30290540, 22 August 2024.

5 https://www.businessinsider.com/elon-musk-20-percent-chance-ai-destroys-humanity-2024-3, 24 August 2024.

6 https://news.uchicago.edu/explainer/origin-life-earth-explained, 22 August 2024.

7 https://www.independent.co.uk/news/uk/hawking-says-computer-virus-is-form-of-life-susan-watts-on-a-manmade-menace-that-mirrors-its-namesake-in-nature-1374137.html, 12 May 2024.

8 It is central to realize that the earth is not a unified organism. Hence, we are not like a virus in a host.

9 https://bigthink.com/the-future/ray-kurzweil-singularity/, 12 May 2024.

10 https://www.bbc.com/news/world-europe-66752264, 12 May 2024.

11 https://www.sciencenews.org/article/satellites-spacex-problem-space-pollution, 12 May 2024.

12 https://www.livescience.com/china-plans-ways-destroy-starlink#:~:text=Chinese%20scientists%20call%20for%20plan%20to%20destroy%20Elon%20Musk's%20Starlink%20satellites,-News&text=Chinese%20military%20researchers%20have%20called,threatens%20China's%20national%20security, 12 May 2024.

13 https://www.newsweek.com/china-fears-spacex-starlink-satellite-national-security-risk-1859322, 12 May 2024.

14 https://www.npr.org/2007/01/19/6923805/chinese-missile-destroys-satellite-in-500-mile-orbit, 12 May 2024.

15 https://www.dezeen.com/2024/05/10/big-moon-3d-printed-data-storage/, 12 May 2024.

16 https://spacenews.com/lonestar-raises-5-million-for-lunar-data-centers/, 12 May 2024.

17 https://pubmed.ncbi.nlm.nih.gov/21204350/, 13 May 2024.

18 https://www.youtube.com/watch?v=PfBG9z8oYCA&list=RDPfBG9z8oYCA&start_radio=1, 3 February 2026.

19 https://www.sciencedirect.com/science/article/pii/S1477388025000131#:~:text=Scientific%20and%20educational%20research%20has,Matthews%20and%20Volpe%2C%202023, 8 July 2025.

20 See also https://bioengineer.org/is-ai-really-creative-new-insights-suggest-creativity-is-subjective/, 8 July 2025.

The Case of Africa, China, and the Meaning of Digital Data

1 This chapter is a slightly revised version of the following article of mine: Stefan Lorenz Sorgner (2022): Africa, China, and the Meaning of Digital Data. *ETHIKA+*, 3(5), 140–150.

2 https://www.forbes.com/sites/wadeshepard/2019/10/03/what-china-is-really-up-to-in-africa/?sh=3d6176e65930, 14 July 2025.

3 Foreign direct investment.

4 https://thediplomat.com/2021/11/the-quiet-china-africa-revolution-chinese-investment/, 14 July 2025.
5 https://www.forbes.com/sites/wadeshepard/2019/10/03/what-china-is-really-up-to-in-africa/?sh=3d6176e65930, 14 July 2025.
6 https://www.universityworldnews.com/post.php?story=20201120110117700, 14 July 2025.
7 https://www.universityworldnews.com/post.php?story=20201120110117700, 14 July 2025.
8 https://www.universityworldnews.com/post.php?story=20201120110117700, 14 July 2025.
9 https://doi.org/10.1016/j.worlddev.2020.105365, 14 July 2025.
10 https://www.iisd.org/articles/chinese-investment-africa-bilateral-trade-decline, 14 July 2025.
11 https://www.ispionline.it/en/pubblicazione/long-trail-china-africa-health-cooperation-aid-trade-and-high-politics-31275, 14 July 2025 (emphasis in original).
12 https://www.ispionline.it/en/pubblicazione/long-trail-china-africa-health-cooperation-aid-trade-and-high-politics-31275, 14 July 2025 (emphasis in original).
13 https://www.nature.com/articles/d41586-018-00927-4, 14 July 2025.

We Have Always Been Cyborgs

1 https://www.nature.com/articles/nature.2016.19136, 8 July 2025.
2 https://di.aerzteblatt.de/int/archive/article/201522/The-relevance-of-irreversible-loss-of-brain-function-as-a-reliable-sign-of-death, 8 July 2025.
3 https://openai.com/index/announcing-the-stargate-project/, 8 July 2025.
4 AI is doing really well in this respect already: https://www.nature.com/articles/s41467-025-59532-5#:~:text=Artificial%20intelligence%20(AI)%20systems%20substantially,trust%20in%20AI%2Ddriven%20decisions, 8 July 2025.
5 https://www.beethovenx-ai.com/, 8 July 2025.
6 https://www.sachsen-fernsehen.de/dresdner-komponist-praesentiert-ki-werk-skills-1557177/, 8 July 2025.

Euro-Transhumanism: *Ex Positivo*

1 This chapter is a revised version of the following article of mine: Stefan Lorenz Sorgner (2026). Euro-Transhumanism. In G. Theiner and A. Malapi-Nelson (Eds) *Technologies of the Future Self: An Ethics for Transhuman Flourishing.* (pp 175–92). Palgrave Macmillan.
2 A great overview: https://scispace.com/pdf/the-history-of-transhumanism-26eov4z8rm.pdf, 8 July 2025.
3 It is interesting to note that Nietzsche, too, stressed that humans cherish an extended healthspan when presenting the, so called, 'last man'. He did not necessarily affirm such a concept, but his analysis reveals that even he recognizes the relevance of an increased healthspan from the perspective of most human beings. However, even the overhuman cherishes health, the great health. Yet, in this case the concept of health is different from the one being cherished by the last human. The great health has to do with the capacity to overcome suffering again and again. In the case of the last human, health has to do with a reduced likelihood of dysfunctionalities coming about in the first place (Sorgner 2025).
4 https://ourworldindata.org/life-expectancy, 21 August 2024.
5 Nietzsche highlighted this insight by stressing the wisdom of the body (Sorgner 2016a; 2016b).

6 Nietzsche talks about overcoming, self-overcoming, and human beings as existences between animals and the overhuman (Blackford 2017; Woodward 2017; Sorgner 2022b).

7 https://www.theatlantic.com/politics/archive/2023/11/peter-thiel-2024-election-politics-investing-life-views/675946/, 21 August 2024.

8 Nietzsche, too, highlights that the soul is merely a part of the body (KSA, Za, 4, 39).

9 Nietzsche regularly highlights correlated insights, and the relevance of encrusted structures, norms, and values for the world we live in (Sorgner 2010c, 241, 196–97; Sorgner 2022b). Bamford rightly stresses that Nietzsche recognizes how difficult it is to move away from traditional values and norms (Bamford 2017).

Euro-Transhumanism: *Ex Negativo*

1 This chapter is a slightly revised version of the following article of mine: Stefean Lorenz Sorgner (2025). What Is Transhumanism Not? A Description Ex Negativo. In:*ANALELE UNIVERSITĂŢII BUCUREŞTI FILOSOFIE*, Vol. LXXIV – Nr. 1, SPECIAL ISSUE: Eternal Returns: Celebrating 180 Years of Nietzschean Thought. In Memoriam George Bondor (pp 107–31).

2 Nietzsche's non-anthropocentric thinking has widely been recognized (Parkes 2005).

3 https://galtonsbox.com/tag/ieet/, 23 July 2024.

4 https://centerforgreatapes.org/orangutan/sandra/, 8 July 2025.

5 Nietzsche's will to power ontology of becoming is a non-dualistic version of naturalism (KSA, NF, 12, 312f; Sorgner 2007).

6 Nietzsche's ontology of becoming can be seen as a reinterpretation of the ontology of becoming as it was presented by Heraclitus (KSA, NF, 12, 211, 312f; Sorgner 2022b).

7 The structures of discrimination of been highlighted by Nietzsche, Foucault (2001), Vattimo (2003), as well as Popper (2003).

8 Some interpreters also regard Nietzsche as an ethical fictionalist (Hussain 2007).

9 Such aspirations can be found in Nietzsche, too (Sorgner 2016a; 2016b).

10 Nietzsche also highlights the relevance of the wisdom of the body with all its idiosyncratic drives, wishes, and longings (KSA, Za, 4, 39).

11 https://ourworldindata.org/, 14 July 2025.

12 https://www.youtube.com/watch?v=QsBT5EQt348, 8 March 2024.

13 https://www.bbc.com/news/blogs-news-from-elsewhere-32050313, 23 July 2024.

14 Tuncel also highlights the relevance of non-duality in Nietzsche's reflections (Tuncel 2014, 91; see also Sorgner 2007).

15 https://www.independent.co.uk/news/uk/hawking-says-computer-virus-is-form-of-life-susan-watts-on-a-manmade-menace-that-mirrors-its-namesake-in-nature-1374137.html, 8 March 2024.

16 There is hardly any philosopher who highlights non-duality more strongly than Nietzsche (Sorgner 2007; 2022b).

17 Nietzsche clearly affirmed perspectivism as his preferred philosophical methodology. It goes along with a version of hermeneutics which is peculiar to Nietzsche (Sorgner 2007).

18 Nietzsche's thinking represents one of the most striking ways of explaining the hermeneutic circle within a non-foundational naturalism (Sorgner 2007). All philosophical judgements are interpretations according to Nietzsche's perspectivism, which again can best be explained on the basis of an ontology of permanent becoming. This non-foundational onto-epistemological grounding is also the most plausible philosophical basis of a philosophically informed transhumanism. My own version of Euro-Transhumanism affirms an analogous philosophical way of thinking.

19 It is also shared by Schopenhauer and Nietzsche (Sorgner 2007; 2022a).

20 https://www.edition-hagia-sophia.de/p/prof-wassilij-a-schipkow-nach-dem-menschen-ideologie-und-propaganda-des-transhumanismus, 4 January 2024.
21 https://pandata.org/transhumanism-and-the-philosophy-of-the-elites/, 8 March 2024.
22 Nietzsche's criticism of Christianity (Prange 2013, 5; Sorgner 2007; 2010c) bears significant similarities to the relationship between transhumanism and Russian orthodoxy.

Religion

1 https://mindmatters.ai/2024/09/fda-lists-neuralinks-blindsight-as-a-breakthrough-device/, 17 July 2024.
2 https://www.bbc.com/news/articles/c1vd097z7rvo, 30 January 2025.
3 https://www.theguardian.com/us-news/2025/jan/20/trump-executive-order-gender-sex, 30 January 2025.
4 https://www.edition-hagia-sophia.de/p/prof-wassilij-a-schipkow-nach-dem-menschen-ideologie-und-propaganda-des-transhumanismus, 30 January 2025.
5 https://www.nytimes.com/2025/01/13/us/politics/steve-bannon-elon-musk.html, 30 January 2025.
6 https://www.theguardian.com/books/2020/jun/05/war-for-eternity-benjamin-r-teitelbaum-starstruck-steve-bannon-dugin-putin-trump, 30 January 2025.
7 https://www.nytimes.com/2022/01/26/world/americas/olavo-de-carvalho-dead.html, 31 January 2025.
8 https://www.newstatesman.com/ideas/2020/10/rise-traditionalists-how-mystical-doctrine-reshaping-right, 31 January 2025.
9 https://www.nationalreview.com/human-exceptionalism/transhumanism-religion-atheists-wesley-j-smith/, 9 July 2025.

Politics

1 This chapter is a revised version of the following article of mine: Stefan Lorenz Sorgner (2025). Final Reflection Liberal Democracies and the Politics of Enhancement. In Hughes, J.J., Umbrello, S., and Calì, C. (Eds) *The Biopolitics of Human Enhancement* (pp 177–90). de Gruyter.
2 https://ourworldindata.org/democracy, 16 June 2024.
3 https://www.edition-hagia-sophia.de/p/prof-wassilij-a-schipkow-nach-dem-menschen-ideologie-und-propaganda-des-transhumanismus, 4 January 2024.
4 https://www.versobooks.com/en-gb/blogs/news/5125-agamben-wtf-or-how-philosophy-failed-the-pandemic, 1 November 2023.
5 https://www.versobooks.com/en-gb/blogs/news/5125-agamben-wtf-or-how-philosophy-failed-the-pandemic, 1 November 2023.
6 It is particularly interesting to note that radical posthumanists still hold on to species type identities rather than focusing on token type identities. They talk about the loss of biodiversity which goes along with an increase of the human population. However, there is no identity between two human beings either. Each human being has a permanently changing idiosyncratic genetic make-up. The more living entities exist on earth, the higher the biodiversity. Biodiversity does not have to be concerned with the quantity of different species, but it could also be concerned with the quantity of living entities.
7 https://dspace.library.uu.nl/handle/1874/386611, 16 June 2024.
8 https://news.cgtn.com/news/3d3d774d79637a4e78457a6333566d54/share_p.html, 16 June 2024.
9 https://transhumanistes.com/on-democratic-transhumanism/, 16 June 2024.
10 https://nickbostrom.com/ethics/infinite.pdf, 16 June 2024.

[11] https://www.theguardian.com/science/2023/jun/14/synthetic-human-embryos-created-in-groundbreaking-advance, 16 June 2024.

[12] https://www.youtube.com/watch?v=n6iOUW523BE&t=128s, 16 June 2024.

Morality

[1] https://www.businessinsider.com/elon-musk-only-chance-of-annihilation-with-ai-2025-2, 23 July 2025.

[2] To actually do what you subscribe to, impulse control is needed. It is a central virtue for Euro-Transhumanism. Hygiene, health, and heart need to be particularly well protected by means of impulse control. The challenge is that a developed prefrontal cortex is needed for impulse control, and you only have this by the age of 25. Hence, it is particularly relevant to promote impulse control when educating a child and teenager.

[3] https://theoxfordculturereview.com/2018/06/10/the-long-view-reconsidering-the-confessions/, 21 August 2024.

[4] https://ourworldindata.org/democracy, 21 August 2024.

[5] https://www.theguardian.com/world/2022/aug/11/china-overtakes-the-us-in-scientific-research-output, 21 August 2024.

[6] https://wng.org/articles/the-puzzle-of-chinas-low-crime-rates-1617299822, 21 August 2024.

Ethics

[1] This is a slightly revised version of the following article of mine: Stefan Lorenz Sorgner (2025). Nietzsche, Dukkha, and Technology: Apology of a Refined Hedonism. In Schussler, Aura Elena (Ed.) *Nietzsche, Critical Posthumanism, and Transhumanism*. Trivent.

[2] https://safetymatters.co.in/2019/05/23/boy-lift-operator-girl-and-finally-ai-aviation-follows-same-trend-mindfly/, 18 July 2024.

[3] https://www.vox.com/future-perfect/411522/self-driving-car-artificial-intelligence-autonomous-vehicle-safety-waymo-google, 9 July 2025.

[4] https://www.scientificamerican.com/article/heres-why-human-sex-is-not-binary/, 14 July 2025.

[5] https://www.bbc.com/news/science-environment-65538866, 14 July 2025.

[6] https://edition.cnn.com/2023/03/24/world/mice-eggs-from-male-cells-scn/index.html#:~:text=Scientists%20create%20mice%20from%20two%20dads%20after%20making%20eggs%20from%20skin%20cells,-By%20Katie%20Hunt&text=Scientists%20have%20created%20mice%20with,significant%20milestone%20in%20reproductive%20biology, 14 July 2025.

[7] https://www.nationalgeographic.com/science/article/early-life-earth-theories, 18 July 2024.

[8] https://pubmed.ncbi.nlm.nih.gov/19092726/, 23 July 2025; https://www.scientificamerican.com/article/when-does-consciousness-arise/, 23 July 2025; https://www.webmd.com/baby/when-can-a-fetus-feel-pain-in-the-womb, 23 July 2025.

[9] https://www.washingtonpost.com/books/2024/03/28/stoicism-ryan-holiday-mark-tuitert/, 18 July 2025.

[10] https://www.catholic.com/qa/what-does-the-term-ex-cathedra-mean-and-where-did-the-catholic-church-come-up-with-it, 18 July 2025.

[11] https://www.tandfonline.com/doi/full/10.1080/0020174X.2016.1258147, 14 July 2025.

[12] https://www.gutenberg.org/cache/epub/21000/pg21000-images.html, 14 July 2025.

If ever I plead with the passing moment,
'Linger a while, you are so fair!'

Then chain me up in close confinement,
Then serving me no more's your care,
Then let the death bell toll my finish,
Then unreluctantly I'll perish,
The clock may stop, the hands fall off,
And time for me be over with!

Goethe, J.W. (2014). *Faust: A Tragedy* (Trans. M. Greenberg). (Para. 1730). Yale University Press.

Education

1 This is a slightly revised version of the following article of mine: Stefan Lorenz Sorgner (2023). What Are the Metahumanities? In Conti, F. and Sorgner, S.L. (Eds) *Humanisms and Beyond: Past, Present, and Future of the Humanities in Liberal Arts Education*. With a Foreword by Franco Pavoncello (pp 145–60). Trivent Publishing.

2 Stefan Lorenz Sorgner (2017). 'Editor's Note', *Journal of Posthuman Studies: Philosophy, Media, Technology*, 1(1), 1–8 ('Reprinted with permission from Penn State University Press.') www.psupress.org/Journals/jnls_JPHS.html, accessed 19 May 2022.

3 Stefan Lorenz Sorgner (2017). 'Editor's Note', *Journal of Posthuman Studies: Philosophy, Media, Technology*, 1(1), 1–8 ('Reprinted with permission from Penn State University Press.') www.psupress.org/Journals/jnls_JPHS.html, accessed 19 May 2022.

4 Stefan Lorenz Sorgner (2017). 'Editor's Note', *Journal of Posthuman Studies: Philosophy, Media, Technology*, 1(1), 1–8 ('Reprinted with permission from Penn State University Press.') www.psupress.org/Journals/jnls_JPHS.html, accessed 19 May 2022.

5 Stefan Lorenz Sorgner (2017). 'Editor's Note', *Journal of Posthuman Studies: Philosophy, Media, Technology*, 1(1), 1–8 ('Reprinted with permission from Penn State University Press.') www.psupress.org/Journals/jnls_JPHS.html, accessed 19 May 2022.

6 Stefan Lorenz Sorgner (2017). 'Editor's Note', *Journal of Posthuman Studies: Philosophy, Media, Technology*, 1(1), 1–8 ('Reprinted with permission from Penn State University Press.') www.psupress.org/Journals/jnls_JPHS.html, accessed 19 May 2022.

7 It seems that the masses live in bubbles of hallucinations. A conscientious use of LLMs helps you push the boundaries of our current state of research. It probably becomes necessary to use them, if you actually want to work on the cutting edge of contemporary research. However, you still need to ask the prompts based on the knowledge in your brain-structure. You have experiences. You turn them into knowledge. You read about topics and connect them to your experiences and your judgements. You ask questions to an LLM and based on your readings and your experiences you manage to separate the helpful responses from the hallucinations. This is what experts can do. If you are not an expert yet, you need to be careful to overtly rely upon LLMs. You need reliable knowledge from experts and feedback from them, as well as the appropriate experiences. A central challenge is being represented by the large group of non-experts who rely on LLMs. They might end up in their own bubble of hallucinations. This is a hermeneutic, Euro-Transhumanist approach to this issue.

8 Stefan Lorenz Sorgner (2018). 'Editor's Note', *Journal of Posthuman Studies: Philosophy, Media, Technology*, 2(2), 123–28. ('Reprinted with permission from Penn State University Press.')

9 www.upress.umn.edu/book-division/series/posthumanities, accessed 15 December 2019.

10 https://www.kcl.ac.uk/ddh/about/about, accessed 15 December 2019.

11 Is the speed of digital progress sustainable? So far, it seems that we are just at the beginning of this development. Once chips move into our bodies, the internet of bodily things

comes about and all aspects of your psychophysiological existence get digitized. This will lift the capacities of LLMs to an entirely new level. In particular, when it comes to biotechnological innovations and insights, we do not even have a glimpse of the potential of LLMs.

12 https://sciencetechnologystudies.journal.fi/, accessed 15 December 2019.

13 Simon Blackburn (2016). *Oxford Dictionary of Philosophy*. Oxford University Press, p 145.

14 Stefan Lorenz Sorgner (2019). 'Editor's Note,' *Journal of Posthuman Studies: Philosophy, Media, Technology*, 3(2), 111–13. ('Reprinted with permission from Penn State University Press.')

15 Evi D. Sampanikou and Jan Stasienko (2021). (Eds) *Posthuman Studies Reader: Core Readings on Transhumanism, Posthumanism and Metahumanism, Posthuman Studies*, volume 2. Schwabe Verlag.

The Case of In Vitro Meat

1 This is a slightly revised version of the following article of mine, which came out in Italian: Stefan Lorenz Sorgner (2025). La carne in vitro: Da una natura profondamente impegnativa a una cultura gloriosamente appagante. In Menin, Marco (Ed.) *Chi può essere mangiato? Filosofia, etica e immaginario della carne coltivata*. Con la collaborazione di Maurizio Balistreri (pp 40–50). Accademia University Press.

2 https://www.osborneclarke.com/insights/italy-bans-lab-grown-meat-violating-eu-procedure, 1 October 2024.

3 https://www.spiegel.de/wirtschaft/italien-verbietet-herstellung-und-verkauf-von-laborfleisch-a-5842c27a-90d0-4578-99a5-50a6728ce9a9, 14 July 2025 (my translation).

4 https://www.osborneclarke.com/insights/italy-bans-lab-grown-meat-violating-eu-procedure, 1 October 2024.

5 https://www.theguardian.com/us-news/2023/jun/21/us-lab-grown-meat-sold-public#:~:text=The%20approval%20means%20that%20the,the%20process%2C%20and%20the%20environment, 1 October 2024.

6 https://www.verbraucherzentrale.de/wissen/lebensmittel/lebensmittelproduktion/clean-meat-ist-laborfleisch-die-zukunft-65071, 1 October 2024 (my translation).

7 https://www.verbraucherzentrale.de/wissen/lebensmittel/lebensmittelproduktion/clean-meat-ist-laborfleisch-die-zukunft-65071, 1 October 2024 (my translation).

8 https://www.vzhh.de/themen/lebensmittel-ernaehrung/fisch-fleisch/laborfleisch-vitro-fleisch-clean-meat-fleisch-der-zukunft, 1 October 2024 (my translation).

9 https://www.theguardian.com/commentisfree/2013/aug/05/worlds-first-cruelty-free-hamburger, 1 October 2024.

10 Zhao, Z., Feng, Q., Yin, Z., Shuang, J., Bai, B., Yu, P., Guo, M., and Zhao, Q. (2017). Red and Processed Meat Consumption and Colorectal Cancer Risk: A Systematic Review and Meta-Analysis. *Oncotarget*, 8, 83306–83314. Retrieved from https://www.oncotarget.com/article/20667/text/, 8 June 2025.

11 https://www.nutrimentumetcurae.com/articles/review/impact-of-artificial-meat-as-novel-food-controversies-and-future-perspectives/, 8 June 2025;
https://tesi.luiss.it/41865/1/760181_PONTECORVI_VALERIA.pdf, 8 June 2025.

12 https://tesi.luiss.it/41865/1/760181_PONTECORVI_VALERIA.pdf, 8 June 2025.

13 https://www.sciencedirect.com/science/article/abs/pii/S0195666318310948?via%3Dihub, 8 June 2025.

14 https://www.technologynetworks.com/cell-science/news/scientists-create-first-adult-mouse-with-two-male-parents-395509, 8 June 2025.

15 The Harmful Beliefs of Photography in Traditional Cultures. *Edge of Humanity Magazine*. https://edgeofhumanity.com/2024/07/30/the-harmful-beliefs-of-photography-in-traditional-cultures/, 8 June 2025.

16 https://ourworldindata.org/life-expectancy?insight=life-expectancy-has-increased-across-the-world#key-insights, 8 June 2025.
17 https://ourworldindata.org/grapher/world-population-in-extreme-poverty-absolute?stackMode=relative, 8 June 2025.

The Case of Surrogacy

1 https://edition.cnn.com/2024/10/19/europe/italy-criminalizes-surrogacy-abroad-intl/index.html#:~:text=Italy%20has%20broadened%20its%20legislation,of%20up%20to%20two%20years, 31 October 2024.
2 https://edition.cnn.com/2024/10/19/europe/italy-criminalizes-surrogacy-abroad-intl/index.html#:~:text=Italy%20has%20broadened%20its%20legislation,of%20up%20to%20two%20years, 31 October 2024.
3 https://www.lto.de/recht/feuilleton/f/peepshow-urteil-bverwg-gute-sitten-verbot-menschenwuerde-rechtsgeschichte, 9 July 2025.
4 https://www.science.org/content/article/fluid-filled-biobag-allows-premature-lambs-develop-outside-womb, 23 July 2025.
5 https://www.bbc.com/news/av/world-latin-america-40634858, 9 July 2025.

The Case of AI, Art, and Creativity

1 https://www.themorgan.org/music-manuscripts-and-printed-music/115399, 9 July 2025.
2 https://www.dailymail.co.uk/news/article-10320437/Pigcasso-piggy-painter-bringing-home-bacon-selling-latest-artwork-20-000.html, 9 July 2025.
3 https://www.dailymail.co.uk/news/article-4719674/Wildlife-photographer-says-famous-monkey-selfie-ruined-him.html, 9 July 2025.
4 https://www.washingtonpost.com/nation/2018/10/26/year-old-developed-code-ai-portrait-that-sold-christies/, 9 July 2025.

Art

1 In this section, you find a summary of selected reflections of mine from Sorgner 2010b.
2 In this section, you find a summary of selected reflections of mine from Sorgner 2010a.
3 All passages from Plotinus' writings were translated by myself.
4 Here, you find a summary of selected reflections of mine from Sorgner 2010b.

The Posthuman Paradigm Shift, Beauty, and Art

1 This chapter is a slightly revised version of the following article of mine: Stefan Lorenz Sorgner (2025). The Posthuman Paradigm-Shift, Beauty, and Art. In: E. Sampanikou and E. Kavakli (Eds) *Digital Wellbeing in the Posthuman Era* (pp 21–9). Trivent Publishing.
2 'Gerade die Autarkie der neuromantischen und symbolistischen Schönheit, ihre Zimperlichkeit jenen gesellschaftlichen Momenten gegenüber, an denen allein Form eine würde, hat sie so rasch konsumfähig gemacht. Sie betrügt dadurch über die Warenwelt, daß sie sie ausspart; das qualifiziert sie als Ware. Ihre latente Warenform hat innerkünstlerisch die Gebilde des l'art pour l'art zu dem Kitsch verurteilt, als der sie heute belächelt werden' (Adorno 1997, 352–3).
3 https://www.theguardian.com/world/2019/nov/07/sandra-orangutan-florida-argentina-buenos-aires, 9 July 2025.
4 'Weil Totalität am Ende die Spannung verschluckt und zur Ideologie sich schickt, wird Homöostase selbst aufgesagt: das ist die Krisis des Schönen und die von Kunst' (Adorno 1997, 85).
5 https://www.patriciapiccinini.net/writing/51, 10 July 2025.

[6] https://faroutmagazine.co.uk/surgery-performance-art/, 10 July 2025.

[7] https://www.labiotech.eu/trends-news/stelarc-ear-art-human-body/, 10 July 2025.

[8] https://www.moma.org/calendar/exhibitions/964, 10 July 2025.

[9] https://news.artnet.com/art-world/agnes-denes-wheatfield-2497536, 10 July 2025.

[10] Donner et al (2016), p 353.

[11] https://www.voanews.com/a/girl-burned-to-death-in-nigeria-after-witchcraft-allegation/2829215.html, 13 January 2026.

[12] https://pulitzercenter.org/stories/branded-and-beaten-nigerian-children-accused-witchcraft-and-murder, 13 January 2026.

The Case of Sex

[1] This chapter is a slightly revised version of the following article of mine: Stefan Lorenz Sorgner (2025). Editor's Note. In *Journal of Posthuman Studies: Philosophy, Media, Technology*, Penn State University Press, 9(2), 123–9.

[2] I am not discussing the relationship between Allen and Farrow in this context.

[3] https://www.independent.co.uk/news/world/americas/elon-musk-daughter-vivian-wilson-b2819120.html, 13 September 2025.

[4] https://www.independent.co.uk/news/world/americas/us-politics/elon-musk-charlie-kirk-assassination-b2824455.html, 13 September 2025.

Bibliography

Friedrich Nietzsche

Sigle of the work edition, siglum of the text, volume number, page number (example: KSA, GT,1, 46).

KSA 1967ff: *Sämtliche Werke – Kritische Studienausgabe in 15 Bänden*. Ed. by G. Colli and M. Montinari, Deutscher Taschenbuch Verlag.

Plotinus

References to Plotinus follow the standard Porphyrian division of the Enneads: Ennead Treatise, Chapter, Lines.

Plotin (1956 ff.). *Plotin's Schriften* (R. Harder, Ed.; continued by R. Beutler and W. Theiler). Wissenschaftliche Buchgesellschaft.

Adorno, T.W. (1997). *Ästhetische Theorie* (Vol. 7). In T.W. Adorno (Ed.), *Gesammelte Schriften in zwanzig Bänden*. Suhrkamp.

Aydin, C. (2017). The posthuman as hollow idol: A Nietzschean critique of human enhancement. *Journal of Medicine and Philosophy*, 42(3), 304–27.

Bamford, R. (2017). Nietzsche on ethical transhumanism. In Y. Tuncel (Ed.), *Nietzsche and Transhumanism* (pp 205–19). Cambridge Scholars Publishing.

Bannon, S.K. (2023). Foreword. In J.J. Allen, *Dark Aeon: Transhumanism and the War Against Humanity*. Simon & Schuster.

Blackburn, S. (2016). *The Oxford Dictionary of Philosophy*. Oxford University Press.

Blackford, R. (2017). Nietzsche, the Übermensch, and transhumanism: Philosophical reflections. In Y. Tuncel (Ed.), *Nietzsche and Transhumanism* (pp 191–204). Cambridge Scholars Publishing.

Blake, W. (1906). *The Marriage of Heaven and Hell*. John W. Luce.

Bostrom, N. (2008). Letter from Utopia. *Studies in Ethics, Law, and Technology*, 2(1), 1–7.

Bostrom, N. (2014). *Superintelligence: Paths, Strategies, Dangers*. Oxford University Press.

Bostrom, N. and Ord, T. (2006). The reversal test: Eliminating status quo bias in applied ethics. *Ethics*, 116(4), 656–79.

Brock, E. (2015). Life is suffering: On Schopenhauer's and Nietzsche's philosophical engagement with suffering. In L.R. dos Santos and K.D. Hay (Eds), *Nietzsche, German Idealism and Its Critics* (pp 189–207). De Gruyter.

Cicero, M.T. (2022). *How to Grieve: An Ancient Guide to the Lost Art of Consolation*. Princeton University Press.

Dinello, D. (2016). The Borg as contagious collectivist techno-totalitarian transhumanists. In W. Irwin (Ed.), *The Ultimate Star Trek and Philosophy: The Search for Socrates* (pp 83–94). Wiley.

Donner, Y., Fortney, K., Calimport, S.R., Pfleger, K., Shah, M., and Betts-LaCroix, J. (2016). Great desire for extended life and health amongst the American public. *Frontiers in Genetics*, 6, 353.

Dugin, A. (2021). Über den Transhumanismus. In W.A. Schipkow (Ed.), *Nach dem Menschen: Ideologie und Propaganda des Transhumanismus* (Philosophia Eurasia, No. 7, pp 133–7). Edition Hagia Sophia.

Etcoff, N. (2011). *Survival of the Prettiest: The Science of Beauty*. Anchor Books.

Flynn, J.R. (2012). *Are We Getting Smarter? Rising IQ in the Twenty-first Century*. Cambridge University Press.

Foucault, M. (2001). Nietzsche, genealogy, history. In J. Richardson and B. Leiter (Eds), *Nietzsche* (pp 341–59). Oxford University Press.

Fukuyama, F. (2004). Transhumanism. *Foreign Policy*, (144), 42–43.

Fuller, S. (2011). *Humanity 2.0: What It Means to Be Human Past, Present and Future*. Palgrave Macmillan.

Fuller, S. (2012). *Preparing for Life in Humanity 2.0*. Palgrave Macmillan.

Fuller, S. and Lipinska, V. (2014). *The Proactionary Imperative: A Foundation for Transhumanism*. Palgrave Macmillan.

Galilei, V. and Palisca, V. (2003). *Dialogue on Ancient and Modern Music*. Yale University Press.

Gemes, K. (2021). The biology of evil: Nietzsche on degeneration (Entartung) and Jewification (Verjüdung). *Journal of Nietzsche Studies*, 52(1), 1–25.

Gödde, G. (2003). Schopenhauer und Nietzsche – zwei gegensätzliche Entwürfe der Lebenskunst: Themenschwerpunkt: Lebenskunst. *Journal für Psychologie*, 11(3), 254–71.

Goethe, J.W. (2014). *Faust: A Tragedy* (Trans. M. Greenberg). Yale University Press.

Haase, R. (1998). *Johannes Kepler's Weltharmonik: Der Mensch im Geflecht von Musik, Mathematik und Astronomie*. Dieterichs Verlag.

Habermas, J. (2004). Freiheit und Determinismus. *Deutsche Zeitschrift für Philosophie*, 52(6), 871–90.

Habermas, J. (2014a). Autopoietische Selbsttransformationen der Menschengattung. In J. Habermas, H.C. Ehalt, U.H.J. Körtner, and P. Kampits, *Biologie und Biotechnologie – Diskurse über eine Optimierung des Menschen* (pp 27–37). Picus Verlag.

Habermas, J. (2014b). *The Future of Human Nature*. John Wiley & Sons.

Habermas, J. (2018). *Die Zukunft der menschlichen Natur: Auf dem Weg zu einer liberalen Eugenik?* Suhrkamp.

Halliwell, S. (2009). *The Aesthetics of Mimesis: Ancient Texts and Modern Problems*. Princeton University Press.

Harris, J. (2011). Moral enhancement and freedom. *Bioethics*, 25(2), 102–11.

Hassan, I. (1977). Prometheus as performer: Toward a posthumanist culture? *The Georgia Review*, 31(4), 830–50.

Hillard, D. (2002). History as a dual process: Nietzsche on exchange and power. *Nietzsche Studien*, 31, 40–56. https://doi.org/10.1515/9783110170740.40

Hughes, J. (2010). Contradictions from the Enlightenment roots of transhumanism. *Journal of Medicine and Philosophy*, 35(6), 622–40.

Hughes, J. (2014). Politics. In R. Ranisch and S.L. Sorgner (Eds), *Post- and Transhumanism: An Introduction* (pp 133–48). Peter Lang.

Hussain, N. (2007). Honest illusion: Valuing for Nietzsche's free spirits. In N. Sinhababu and B. Leiter (Eds), *Nietzsche and Morality* (pp 157–91). Clarendon Press.

Huxley, J. (1951). Knowledge, morality, and destiny. *Psychiatry*, 14(2), 129–40.

Kain, P.J. (2007). Nietzsche, eternal recurrence, and the horror of existence. *Journal of Nietzsche Studies*, 33, 49–63. https://doi.org/10.1353/nie.2007.0007

Kurzweil, R. (2005). *The Singularity Is Near*. Viking.

Kurzweil, R. (2014). The Singularity Is Near. In R. Sandler (Ed.), *Ethics and Emerging Technologies* (pp 393–406). Palgrave Macmillan.

Leiter, B. (1992). Nietzsche and aestheticism. *Journal of the History of Philosophy*, 30(2), 275–90. https://doi.org/10.1353/hph.1992.0020

Mackie, J.L. (1990). *Ethics: Inventing Right and Wrong*. Penguin.

More, M. and Vita-More, N. (Eds) (2013). *The Transhumanist Reader: Classical and Contemporary Essays on the Science, Technology, and Philosophy of the Human Future*. Wiley-Blackwell.

Parkes, G. (2005). Nietzsche's environmental philosophy: A trans-European perspective. *Environmental Ethics*, 27(1), 77–91.

Pearce, D. (2015). Negative utilitarianism: Why be negative? In *The Hedonistic Imperative*. Retrieved from https://www.hedweb.com/negutil.htm

Pearce, D. (2021). Hedonist Imperative. In E.D. Sampanikou and J. Stasienko (Eds), *Posthuman Studies Reader: Core Readings on Transhumanism, Posthumanism and Metahumanism* (pp 57–64). Schwabe Verlag.

Persson, I. and Savulescu, J. (2008). The peril of cognitive enhancement and the urgent imperative to enhance the moral character of humanity. *Journal of Applied Philosophy*, 25, 162–77.

Persson, I. and Savulescu, J. (2012). *Unfit for the Future: The Need for Moral Enhancement*. Oxford University Press.

Persson, I. and Savulescu, J. (2013). Getting moral enhancement right: The desirability of moral bioenhancement. *Bioethics*, 27(3), 124–31.

Pinker, S. (2011). *The Better Angels of Our Nature: Why Violence Has Declined*. Viking.

Popper, K.R. (2003). *Die offene Gesellschaft und ihre Feinde: Der Zauber Platons* (Vol. 1). Mohr Siebeck.

Prange, M. (2013). *Nietzsche, Wagner, Europe* (Vol. 61). Walter de Gruyter.

Ranisch, R. and Sorgner, S.L. (Eds) (2014). Introducing post- and transhumanism. In *Post- and Transhumanism: An Introduction* (pp 7–28). Peter Lang.

Rothblatt, M. (2011). *From Transgender to Transhuman: A Manifesto on the Freedom of Form*. Martine Rothblatt.

Safronov, D. (2022). Of last men and the ends of history: Nietzsche contra Fukuyama. In M.A. Ruehl and C. Schubert (Eds), *Nietzsches Perspektiven des Politischen* (pp 355–68). De Gruyter.

Sampanikou, E. and Stasienko, J. (Eds) (2021). *Posthuman Studies Reader*. Schwabe Verlag.

Sandberg, A. (2013). Morphological freedom: Why we not just want it but need it. In M. More and N. Vita-More (Eds), *The Transhumanist Reader* (pp 56–64). Wiley-Blackwell.

Sandel, M.J. (2007). *The Case Against Perfection: Ethics in the Age of Genetic Engineering*. Harvard University Press.

Savulescu, J. (2001). Procreative beneficence: Why we should select the best children. *Bioethics*, 15(5–6), 413–26.

Schussler, A.E. and Balistreri, M. (Eds) (2024). *Metahumanism, Euro-Transhumanism, and Sorgner's Philosophy: Technology, Ethics, Art*. Trivent Publishing.

Sier, K. (2010). Platon. In S.L. Sorgner and M. Schramm (Eds), *Musik in der antiken Philosophie: Eine Einführung* (pp 123–66). Königshausen & Neumann.

Sloterdijk, P. (2001). *Nicht gerettet: Versuche nach Heidegger*. Suhrkamp.

Sorgner, S.L. (2007). *Metaphysics without Truth: On the Importance of Consistency within Nietzsche's Philosophy* (2nd edn). Marquette University Press.

Sorgner, S.L. (2009). Nietzsche, the Overhuman, and transhumanism. *Journal of Evolution and Technology*, 20(1), 29–42.

Sorgner, S.L. (2010a). Plotins Musikphilosophie. In S.L. Sorgner and M. Schramm (Eds), *Musik in der antiken Philosophie: Eine Einführung* (pp 275–94). Königshausen & Neumanny.

Sorgner, S.L. (2010b). Einige Überlegungen zur antiken und modernen Musikphilosophie. In S.L. Sorgner and M. Schramm (Eds), *Musik in der antiken Philosophie: Eine Einführung* (pp 15–32). Königshausen & Neumann.

Sorgner, S.L. (2010c). Menschenwürde nach Nietzsche: Die Geschichte eines Begriffs. Wissenschaftliche Buchgesellschaft.

Sorgner, S.L. (2016a). Perfecting human beings: From Kant and Nietzsche to trans- and posthumanism. *Trans-Humanities*, 9(1), 41–61.

Sorgner, S.L. (2016b). From Nietzsche's Overhuman to the posthuman of transhumanism: Transcultural discourses. *The Journal of English Language & Literature*, 62(2), 163–76.

Sorgner, S.L. (2017). Editor's note. *Journal of Posthuman Studies: Philosophy, Media, Technology*, 1(1), 1–8.

Sorgner, S.L. (2018a). *Schöner neuer Mensch*. Suhrkamp.

Sorgner, S.L. (2018b). Editor's note. *Journal of Posthuman Studies: Philosophy, Media, Technology*, 2(2), 123–8.

Sorgner, S.L. (2019a). *Übermensch: Plädoyer für einen Nietzscheanischen Transhumanismus* (Schwabe Reflexe 59). Schwabe Verlag.

Sorgner, S.L. (2019b). Editor's note. *Journal of Posthuman Studies: Philosophy, Media, Technology*, 3(2), 111–13.

Sorgner, S.L. (2019c). Android Data: Eigentum oder Träger von Rechten? In M.C. Bauer (Ed.), *Neue Welten – Star Trek as Humanistic Utopia?* (pp 239–72). Springer.

Sorgner, S.L. (2020). *On Transhumanism* (Trans. S. Hawkins). Penn State University Press: University Park.

Sorgner, S.L. (2021). Adage of Metahumanism. In E.D. Sampanikou and J. Stasienko (Eds), *Posthuman Studies Reader: Core Readings on Meta-, Post- and Transhumanism* (pp 305–8). Schwabe Verlag.

Sorgner, S.L. (2022a). *We Have Always Been Cyborgs: Digital Data, Gene Technologies, and an Ethics of Transhumanism*. Bristol University Press.

Sorgner, S.L. (2022b). *Philosophy of Posthuman Art*. Schwabe Verlag.

Sorgner, S.L. (2022c). On constructive and destructive interferences in the life and in art worlds: A reply. *Deliberatio: Studies in Contemporary Philosophical Challenges*, 2(2), 170–86.

Sorgner, S.L. (2025). Glocal challenges of Euro-Transhumanism. In N.A. Rojas Cortés (Ed.), *Transhumanism and Latin America: Symposium on Sorgner's 'We Have Always Been Cyborgs'*. Trivent Publishing.

Sorgner, S.L. and von Becker, P. (2023). *Transhumanismus*. Westend Verlag.

Thomas, A. (2024). *The Politics and Ethics of Transhumanism*. Bristol University Press.

Tomlinson, G. (1993). *Music in Renaissance Magic: Toward a Historiography of Others*. University of Chicago Press.

Tuncel, Y. (2014). Nietzsche. In R. Ranisch and S.L. Sorgner (Eds), *Post- and Transhumanism: An Introduction* (pp 83–100). Peter Lang.

Tuncel, Y. (Ed.). (2017) *Nietzsche and Transhumanism*. Cambridge Scholars Press.

Varchi, B., Bätschmann, O., and Weddigen, T. (2013). *Paragone: Rangstreit der Künste: Italienisch und Deutsch*. Wissenschaftliche Buchgesellschaft.

Varela, F.J., Thompson, E., and Rosch, E. (2017). *The Embodied Mind: Cognitive Science and Human Experience*. MIT Press.

Vattimo, G. (2003). *Glauben – Philosophieren* (Trans. C. Schultz). Reclam Verlag.

Walker, D.P. (2000). *Spiritual and Demonic Magic: From Ficino to Campanella* (Vol. 22). Penn State Press.

Woodward, A. (2017). Postmodern reflections on the Nietzsche and transhumanism exchange. In Y. Tuncel (Ed.), *Nietzsche and Transhumanism* (pp 231–47). Cambridge Scholars Publishing.

Index

1984 (Orwell) 93
2001: A Space Odyssey 96

A

Abramović, M. 180
academia 17, 58, 110
accelerationism 16
actuators 52–5, 57
Adorno, T. 173, 177, 180
Africa 45–50
afterworld 127
Agamben, G. 105
ageing 12, 13–14, 98
see also healthspans, increasing
agency, human versus artificial 53, 57–8
AGI (artificial general intelligence) 34, 35, 55, 68–9, 164
AI (artificial intelligence) 15, 34–44, 53, 57, 162–4
see also AGI, superintelligence
aletheism 26, 86–7, 100–1, 140
algorithms 32, 34, 42, 53, 55–6, 58, 72, 83
America PAC 11
androids 36
Anglo-American traditions 9, 14, 18, 19, 63, 66, 94, 109
anthropocentrism 30–1, 75, 140, 178, 179–80
anti-ageing research 13–14
antibiotics 31, 154, 157
anti-democracy 21–2
anti-libertarianism 11
antinatalism 16
anti-perfectionism 11
anti-pluralism 28
anti-Semitism 12, 99, 111
anti-transhumanism 12, 99
anti-utilitarianism 11
anti-utopianism 11
Apollo 171–2
Apollo of the Belvedere 179–80
Aristotle 91, 97, 102, 160, 172
art 96, 133, 134, 148, 162–4, 165–76, 177–86
artificial agents 53
artificial consciousness 36
astrophysics 3
atheism 100, 101
Augustine, Saint 102, 119
authoritarianism 32, 57, 82–3, 91, 104
autobiography 119
automation 124
autonomy 5, 66, 102, 112, 113, 120, 187–8
avatars 26, 52, 53, 78

B

Bannon, S. 12, 99, 111
base hedonism 128, 134, 135
beauty 168–9, 173–5, 177–86, 189
becoming, ontologies of 18, 51, 71, 78, 86, 87, 100, 107–8, 121, 173
Bentham, J. 122
Bergamini, D. 151
Berlin style 128–9
Big Bang Theory 96
big data analysis 40, 82
Big Tech 17–18, 22, 83
bioart 96
bioconservatism 27, 110, 111, 112, 113
bioliberalism 19, 110
biopolitics 105
bioprinting 125, 156, 160
biotechnology 13
'Blindsight' 96
bodily integrity 26, 52, 78
body modification 89–90
body-centric traditions 70
body-soul 94, 108, 139, 143
Bolsonaro, J. 12, 99
Borg 36, 38
Bostrom, N. ix, 9, 10, 16, 20, 21–2, 23, 38, 62, 63, 64, 79, 87, 89, 93, 94, 110, 156
Braidotti, R. 16, 106
brain-computer interfaces 27, 40, 41, 52–5, 72, 73, 96, 118
brains 41, 54, 182
Bratton, B. 105
Brave New World (Huxley) 10, 93
Brazil 12, 99
Brown, D. 11, 95

Buddhism 14, 69, 89, 100, 125–6, 130, 155, 183
Buterin, V. 13, 22–3, 98

C

Camerata Fiorentina 170–1
cancel culture 183, 185–6, 193
carbon dioxide emissions 31
carbon-based transhumanism 15, 67–8, 72
Carvalho, O. de 12, 99
Casalis, P. 17
Catholicism 3, 14, 51, 72, 94, 97, 101–3, 131–2, 156, 187
chastity 187
ChatGPT 41, 42, 43, 56, 164
Chiang Mai, Thailand 13, 95
China 12–13, 30–2, 40, 45–50, 56, 81–3, 106, 122, 155
Christianity 94, 101, 102, 104, 125–6, 132, 165, 172
Cicero 130, 140–1, 143
Clarke, A.C. 96
classic transhumanism
  anthropocentrism 75
  concepts of the good 62
  differences between CT (classic transhumanism) and ET (Euro-Transhumanism) ix, 8, 9, 14–15, 16, 20, 64–71, 75–92
  history of 62–3
  libertarianism 109, 110
  naïve hedonism 13
  overview 10
  politics 98
  religion 94, 97
  sexuality 193
climate change 80
  see also global warming
club 27 128–9
cognitive capacity enhancement 117, 118–19
communism 110
compassion 117
computer simulations 97–8
consciousness 36–9, 52, 54, 57–8, 68, 85, 114, 126–7
consent 188–93
consumerism 180
copyright 41–2, 163
cosmic singularity 61, 67
COVID-19 48, 105
creativity 42–3, 59, 133, 162–4
critical posthumanism
  aesthetics 180
  autonomy 188
  consent 191–2
  discrimination 97
  education 138
  evolution of Homo Sapiens 26
  liberal democracies 107, 110
  metahumanism 110
  mind uploading 139
  monsters 185
  natural law 187
  personhood 76
  versus radical posthumanism 105
  subject-object distinction 107, 109
  versus transhumanism 16, 18, 27, 61
cryonic preservation 85–6
cryptocurrency 13, 98
cults 88, 95
cyborgs 15, 25, 51–60, 68, 73, 94, 116–18, 178, 180

D

Dante 9, 61
Dark Aeon (Bannon, 2023) 12, 99
Darwin, C. 51, 76, 93, 101, 127, 138, 141, 155, 177, 178
death 3, 54, 64, 126–7, 129, 131
decelerationism 16
dehumanization 23
Deleuze, G. 18, 108
democracy 22, 82, 175
demons 165, 169, 170, 172–3
Denes, A. 181
Descartes, R. 26, 69
deserts 4, 5, 100–1
desires 14, 79
digital agents 53–4
digital consciousness 35–9, 41, 43
digital data 31–2, 39–41, 45–50, 56–9, 82–3, 122
digital humanities 144–5, 147
digital literacy 144, 145, 147, 149
digital panopticon 68–9, 122
dignity 51, 52, 82, 111, 112, 140–1, 159, 177
Dinello, D. 38
discrimination 26, 86, 97, 178, 187
diversity 64–5, 90, 106
Divine Comedy (Dante) 9, 61
divine spark 73, 94, 113, 140, 177
domination structures 26
drives 4, 5, 64, 71, 90, 100, 120–1, 135, 191
drugs 128–9
dualist ontologies 18–19, 26, 68, 70, 73, 85, 94–7, 138–9, 143
Dugin, A. 12, 28, 29, 91, 99, 100, 104
Dukkha 14, 65, 69, 100, 125–36, 175
dystopia 93, 96

E

Edge City Lanna 13
education 138–50
educational freedom 71
Elmauer Rede (Sloterdijk) 5, 27

embodied AI 35, 53, 54, 68–9
embodied minds 70, 77, 78
emotions 37, 38, 39, 43, 57–8, 68, 126
enhancement technologies
  critical posthumanism 138–9
  morality 116
  politics 109–15
  in vitro meat 155
  see also cyborgs; healthspans, increasing
Enlightenment 15, 19, 20, 57, 69–70, 76, 83, 101, 102, 103, 138
entrepreneurs 22, 83
Epicureanism 127, 129
epigenetics 112, 141, 142
equality 82, 111–15
Esfandiary, F.M. *see* FM-2030
essentialist understandings of human nature 19, 51, 77–9, 108, 113–14, 139–41, 177, 191
Etcoff, N. 181–2, 189, 190, 192
Ethereum 13, 23, 98
ethical challenges 66
ethics courses 147–8, 149
ethics of AI 38, 42
eugenics 10, 23, 27, 29, 72, 84, 155
Eurasian Orthodoxy 28, 91, 104
euthanasia 52–3
evolution
  AI 37
  art 178
  brave new humans 25, 27–8
  CT (classic transhumanism) versus ET (Euro-Transhumanism) 67–8
  cyborgs 51
  entanglements of transhumanism 18
  language development 25, 51, 52, 73, 94, 117
  metahumanities 142
  and reason 52
  sex 189, 191
  silicon-based lifeforms 86
  in transhumanist sense 9
Ewha Womans University, Seoul 148–9
extinction 22, 25, 27, 105, 110, 116

F

Faust (Goethe) 3, 136
Fibonacci sequence 167–8
Ficino, M. 165, 169, 170–3
fiducia 7, 95, 135, 136, 137
films 95–6
Florence style 132–4
Flynn, J.R. 118
FM-2030 10, 16, 62–3, 93
Foucault, M. 18, 108, 128
Frankfurt school 173, 180
freedom 57, 71, 80–1, 84, 85, 89, 109, 120
Freud, S. 51, 101, 119, 141, 177, 178, 187, 191
Fukuyama, F. 11, 12, 31, 99, 110, 111, 112
Fuller, S. 19, 31, 75, 76, 110
fundamentalism 104, 187
Future of Humanity Institute (FHI) 10, 23, 93, 97
Future Studies/Futurology 146–8

G

Gattaca 96
GenAI (generative AI) 41–3, 59
gender identities 91, 96–100
gene technologies 29, 72, 73, 96, 113, 118, 125, 183
General Data Protection Regulation 32, 57, 82–3
genetic mutations 25
genetic testing 84
genetics 58
Germany 27
Glass, P. 177, 181, 182, 183, 185–6
global anti-transhumanist coalition 12
global government 116
global warming 17, 26–7, 29–30, 31, 80, 81, 155, 157
glocalism 9, 63
Goethe, J.W. von 3, 11, 95, 108, 136
golden mean 167, 179–80, 183, 185
good, concepts of the
  ex negativo ET 79, 84
  ex positivo ET 62, 64, 71, 72
  internet panopticon 122
  politics 104
  radical plurality of the good 14, 71, 79, 119–20
  twisted concepts of goodness 65
  virtues 120–1
Good, I.J. 34
grandezza 7, 95, 135, 136, 137
Grey, A. de 13
Grimalda, G.F. 17
Grochowska, J. 63, 96

H

Habermas, J. 11, 27, 28, 77–8, 88, 89, 95, 110, 111
Harris, J. 110
Hassan, I. 16
Hawking, S. 36, 37, 86
Hayles, N. K. 16, 109
healthspans, increasing
  alternative health systems 184
  CT versus ET 67
  digital data 32
  entrepreneurs 22
  great health 133
  living a good life 65
  meat eating 156
  non-perfectionism 79

not the same as lifespan 61
personal data 83
and power 133
quality of life 47, 61–2, 65
SENS Research Foundation 11–12, 13, 23, 98
symmetry 181–3
technologies 73
transhumanism 11–12, 13, 14, 15, 20
virtues 120
in vitro meat 153–4
see also quality of life
hedonism 6–7, 13, 94, 127, 128–9, 134, 135
Hegel, G.W.F. 119
Helbig, S. 58–9, 63, 96, 163, 177, 181, 182, 183, 185–6
Heraclitus 3, 4, 11, 63, 71, 77, 95, 136
hermeneutic ethics 14, 66, 101
heteronormativity 21, 22, 89–91, 97, 102, 115
Homo sapiens 25–6
Hughes, J. 63, 76, 109, 110
human exceptionalism 139–41, 178
human flourishing 15, 27, 80, 91, 95
human intelligence 34, 55, 118
humanism 21, 26, 61, 97, 127, 139, 178, 179
humanities 58–9, 140–9
Humanity Plus 23
human/'other' boundaries 138
Huxley, A. 10, 93
Huxley, A.F. 10, 93
Huxley, J. 9–10, 16, 20, 61, 62, 93, 109
Huxley, T.H. 10, 93
hyper-humanism 20–1, 28, 30–1, 64, 85, 139
Hyper-Renaissance 14

I

identity over time 113
immateriality 19, 26, 51–2, 77–8, 85, 94, 113, 127, 143, 177
immortality 13, 15, 28, 61, 67, 68, 98, 139
impulse control 117, 121
in vitro meat 151–8
incest 84
India 30, 31, 81, 155
indigenous culture 184–5
inequality gaps 82, 111–15
Inferno (Brown, 2013) 11, 95
informed judgements 18
injustice 31–2
innovation 41–3, 59
Institute for Ethics and Emerging Technologies 31
Institute for the World Economy (IfW) 17
intentionality 37, 38, 43, 53, 54, 57–8, 68, 163–4
internet panopticon 116, 117–18, 121–3
is-ought fallacy 107
Istvan, Z. 82
Italy 151, 152, 153, 159

J

John Cabot University, Rome 148–9
Judeo-Christian tradition 108, 114, 188

K

Kac, E. 96, 177, 181, 182, 186
Kant, I. 26, 51, 69, 119, 159, 183
Kepler, J. 166, 168
Kirill I, Patriarch of Moscow 12, 91, 99, 104
Kurzweil, R. 38, 61, 87

L

language, development of 25, 51, 52, 73, 94, 117
last human being 129, 133
left-wing politics 29, 105
LGBTQIA+ movement 12, 98–9, 104–5
liberal democracies
critics of 104–7
defenders of 107–11
see also social liberal democracies
liberalism 29, 82, 84, 120
libertarianism 10, 15, 20, 29, 56–7, 66–7, 71, 82, 109–10
life 37–8, 86, 125
lifespan increases 12–13, 62
see also healthspans, increasing
light fictions 174
Locke, J. 40
logocentrism 69–70, 76–7, 179
Lonestar Data Holding Inc 40
longtermism 18
love vii, viii, 96–100, 102, 103

M

Mackie, J.L. 65
magic 184–6
Marsyas 171
mathematics 3, 87, 166–7, 169, 174
meaningfulness ix, 6, 69
meat eating 30, 31, 80–1, 151–8
media portrayals of transhumanism 20, 95–6
metahumanism 110
metahumanities 141–8, 178
metamorphosis 4
metaphor 4
metaverse 26, 78
Methuselah Foundation 11, 23
Milan style 130–1
mind uploading
brave new humans 27
classic transhumanism 15
critical posthumanism 139
CT versus ET 67, 68, 85

entanglements of transhumanism 20
ex positivo ET 61
and suffering 65
mind-body dualism 70, 139
mindfulness 5, 6, 7, 117, 121, 155
mirror test 54, 76, 112, 114, 118–19
money 6
monsters 175, 179, 180, 181, 182, 185
Moore's Law 38
moral dilemmas 66
moral enhancement 116–23
moral respect 111, 113–14
More, M. 10, 16, 20, 27, 63, 93, 94, 97
morphological freedom 14, 47, 62, 71, 80, 89, 188
music 58–9, 165–76
Musk, E. ix, 9, 10, 11, 17–18, 22–3, 27, 36, 39, 44, 63, 82, 93, 94, 96, 97, 110, 116, 193

N

naïve hedonism 6–7, 13, 94, 127, 134
Nancy, J.-L. 105
nanobots 36, 39
Naples style 134–5
natural law 72, 89, 91, 97, 102, 156, 160, 187
naturalist ontologies 70, 71, 76–8, 88
naturalistic anthropology 27
nature/culture 156–7, 160
negative freedom 19, 57, 71, 109
neo-colonialism 65, 66, 108, 117
Neo-Platonics 165, 172
Neuralink 27, 38, 39, 40, 41, 44, 72, 96
neurogenesis 41
neurology 41, 182
Nietzsche, F. 3, 4, 11, 15, 20, 28, 51, 63, 69, 70–4, 76, 92, 95, 101, 119, 127, 129, 130, 131, 132, 134, 135, 136, 141, 155, 177, 178, 191
nihilism 4, 100–1, 183
non-dualism 18–19, 67, 71, 85–6, 108, 138–9, 142, 155, 157, 178
non-human animals 31, 90, 140, 162, 178
non-human personhood 31, 75, 107, 111–12, 157
numerical ratios 167, 174, 179, 183

O

On Nature (Heraclitus) 3
O'Neill, J. 12, 97, 98
OpenAI 163
opera 171
opinions versus informed judgements 18
Orangutans 76, 178
original transhumanism 10, 93, 97
Orlan 180
Orthodox church 12, 98–9, 104, 187
Orwell, G. 93
ourworldindata.org 30, 81, 104, 119
overhumans 113, 133
overpopulation 26–7, 29–30, 81
oxytocin viii

P

pain 114, 125–9, 133, 135
see also pleasure–pains; suffering
Palantir 11
Pan vii, viii
panopticon 68–9, 116, 117–18, 121–3
papal infallibility 131–2
paragone 171–2
paternalism 5, 14, 27, 29, 49, 80–1, 101, 106–7, 120, 173
Pearce, D. 10, 87
perfectionism 10, 20, 21, 64, 72, 79–80, 97
personal data 32, 40–1, 82–3, 84, 88
personal flourishing 91, 97, 120
personal rights 106
personhood 31, 52, 75–6, 107, 111–15, 178
perspectivism 4–5, 14, 65–6, 86–7, 100–2, 108, 188–9
Persson, I. 116
pessimism 65, 89
philosophy 3–5, 11, 18, 87, 89, 100–1, 108, 113, 134, 169–70
physics 3
Piccinini, P. 96, 180
Pigcasso 162
Pinker, S. 117, 118
Plato 3, 26, 51, 66, 73, 132, 139, 141, 165, 170, 171, 172, 174, 187
pleasure–pains 7, 65, 69, 131, 134, 135
Plotinus 168–70
pluralism 12, 28, 29, 89, 91
political subjectivity 107
politics 20, 98, 104–15
Pope Leo XIV 13
population growth 30, 81
pop-up cities 13, 95
Portrait of Edmond Belamy 163
posthumanism
art 177
beauty 180, 181
entanglements of transhumanism 18–19
on liberal democracies 105, 108
music and becoming 173–5
paradigm shift ix, 95, 97, 138–9, 148–9, 155, 157, 165, 173, 177–9
plurality of meanings 138
posthuman hedonism 127, 134
posthuman studies 142–6
posthuman virtues 117–18, 119–21
seriousness 174–5
versus transhumanism 8, 9, 72
in transhumanist sense 109
see also critical posthumanism

postmodernism ix, 105, 175
power 6, 131–2, 133, 134, 135, 188, 191
privacy 57, 82
procreation 27, 30, 84
see also reproduction
professionalization 124–5
profit motives 83
Prometheus viii
pronatalism 16
property theory 40, 84, 159
Proteus vii, viii
public health care 67, 71, 83, 84
Putin, V. 12
Pythagorean musical tradition 165, 166–8

Q

quality of life
Africa 47
CT versus ET 71–2
digital data 83
digital information flows 82
ex positivo ET 61–2
population growth 81
radical plurality of the good 14, 71, 79, 119–20
symmetry 182–3
technologies 27, 73, 88–9
transhumanism 19–20
virtues 120–1
see also healthspans, increasing

R

racism 21, 22, 26, 86, 97, 115, 178
radical plurality of the good 14, 71, 79, 119–20
radical posthumanism 16, 26, 105–6
radical transhumanism 27
raffinatezza 7, 95, 135, 136, 137
Ranisch, R. 18, 27, 108, 177
rationality 17, 18, 25, 68, 70, 94, 113, 140, 159, 179
reason 51, 52, 70, 77, 78, 94, 169, 172, 175
relational ethics 107, 135
relational subjectivity 106
religion 93–103, 104, 106, 110, 187
Renaissance 64, 72, 79, 97, 171
reproduction 72, 80, 84, 89–90, 97, 124–5, 156, 159–61, 191
reproductive freedom 71
Republic (Plato) 3
Researcher, The (Casalis, 2024) 17
Riefenstahl, L. 181, 182
rights-based approaches 107, 109, 112
robots 35–6
Rome style 131–2
Roser, M. 30, 81
Rothblatt, M. 63, 90, 97, 99, 190, 193
'Rules for the Human Zoo' (Sloterdijk) 27
Russia 12, 28, 91, 98–9
Russian Science Foundation (RSF) 12

S

sanctions 84, 118, 122
Sandel, M.J. 110, 111
satellites 40
Savulescu, J. 64, 110, 116, 152
Schopenhauer, A. 95, 127, 130, 191
science 3, 28–9, 87, 109, 145, 148
science technology studies 145–8
self-care 6–7
self-consciousness 37, 38, 43, 52, 54, 58, 68, 114, 121, 126
self-contradictions 17
self-control 5–6, 7, 117, 121, 133
SENS Research Foundation 11–12, 13, 23, 98
sex 187–93
sex and gender 96–100
sex workers 159
sexism 21, 26, 86, 97, 115, 140, 178
sexuality 13, 72, 102, 125, 156, 161, 187–93
silicon-based entities 37–8, 41, 52, 53, 54, 67–8, 72, 86, 164
silicon-based transhumanism 49
silicon-carbon interfaces 52–3
simulation hypothesis 10, 23
Singer, P. 18, 76, 152
singletons 22
singularitarianism 18, 68
slavery 112, 130
Sloterdijk, P. 5, 27
smartphones 25–6, 52, 78
social credit 32, 45, 106, 122
social liberal democracies 15, 16, 29, 66–7, 71, 82, 109, 110
social media 120
social networks 80, 106, 135
soft naturalism 77–8
Sora 41, 43, 163
Sorgner, S.L. 9, 10, 14, 18, 20, 25, 27, 35, 37, 51, 52, 61, 71, 73, 76, 78, 79, 83, 85, 86, 87, 93, 106, 107, 108, 109, 110, 111, 116, 117, 120, 121, 122, 131, 138–9, 142–4, 146, 155, 156, 173, 174, 177, 178
souls 51, 71, 85, 94, 108, 112, 139, 170, 177
space 40, 98
speciesism 26, 31, 86, 97, 110, 140, 178
Spinoza, B. 11, 63, 95, 127, 155
spirit 78, 139, 165, 166, 170
Star Trek 35–6, 38
Stargate 56
Starlink 38, 39–40, 41, 44
Stelarc 96, 180
Stoics 26, 130–1, 141
storytelling 65
strong AI 35

subjectivity 106, 114
subject-object distinction 105, 107, 109, 110, 111, 112, 143
suffering 15, 52, 65, 69, 89, 114, 125–36, 157
superintelligence 15, 34–44, 55, 68–9, 94, 163–4
surrogacy 159–61
sustainability 29, 47, 80
symmetry 168–9, 179–81, 189, 191
synthetic embryos 112

T

tax 80, 83
Taylor Greene, M. 105
tech entrepreneurs 17–18
technological singularity 38
technologies
  automation 124
  and the definition of transhumanism 19
  digital humanities 147
  humanities 144
  language as 73
  quality of life improvements 71–3, 88–9
  reproduction 72
  see also enhancement technologies
technoprogressive transhumanists 109, 110
Teiresias 90–1
TESCREAL theory 17–18, 22
Thiel, P. 9, 11, 22–3, 63, 71, 82, 94, 97–8, 110
Third Fellowship 98
Third Reich 70, 72, 118, 155, 182
three-parent children 90, 161
Thus Spoke Zarathustra (Nietzsche) 3
Tomlinson, G. 165, 170, 172
totalitarianism 23, 28–9, 80, 81–5, 101, 105, 182
traditionalism 12, 99, 152–3
Transcendence 96
transhumanism
  in the 21st century 11
  compared to critical posthumanism 18–19
  definition 9
  entanglements of 16–24
  future of 13–14
  history of 9–10
  politics today 11–13
  varieties of 10–11
Trump, D. 11–12, 98, 99
trust viii, 135, 189, 190
truth 26, 77, 78, 87, 91, 100–1, 102, 191
truthfulness 5, 6, 7, 117, 120
Tuncel, Y. 63, 70–1
Turin style 129–30

U

Übermensch 113
universally valid norms 16, 51, 63, 65, 66, 173, 181, 183, 185–6
USA 12, 32, 56, 82, 83
utilitarianism 10, 14, 20, 28, 64, 66, 84, 110, 130
utopias 10, 14, 20, 64, 87–8, 94–6, 106

V

values and norms 5, 65–6, 78–9, 184
vampire capitalism 29–30
Vance, J.D. 11
Vattimo, G. 66, 101, 102, 103
violence 79, 91, 101, 118
virtue ethics 10, 130
virtues 6, 119–21
viruses 37, 86
Vita-More, N. 10, 16, 27, 63, 93, 97
Vitruvian Man 79, 179
Vom Lärm der Welt (On the Noise of the World) 96
von Leys, T. 96
Vorontsova, M. 12
voting rights 112

W

Wagner, R. 11, 95, 185
Walker, D.P. 165, 170, 172
We (Zamyatin) 93
'weak AI' 34–5, 43
welfare state liberalism 20
witchcraft 184–6
'woke' 22
World Humanities Forum 142
World Transhumanist Association 10, 23

X

Xi, J. 12–13

Z

Zamyatin, Y.I. 93
Zuzala pop-up city 13